NUMBERS AT WORK

by John Gillespie

Making numbers work for you

NATIONAL EXTENSION COLLEGE CAMBRIDGE

ISBN: 0 86082 155 2.
First edition 1980.
Designed and Produced by Peter Hall.
Photographs by Michael Manni.
Cover Photograph by Reeve Photography.
Illustrations by Peter Welford.

Printed by NEC Print, 18 Brooklands Avenue, Cambridge CB2 2HN.
Published by the National Extension College, Cambridge.
    The National Extension College is an adult teaching body providing education through correspondence courses, publishing, tapes, slides and kits. As a non-profit distributing company, registered as a charity, all monies received by NEC are used to extend and improve the range of services provided. Write to the Course Texts Department for a catalogue of materials available.

Author's note
I would like to thank Dr. Bob Laxton of the University of Nottingham, Mr. Reg Slack of Trent Polytechnic and Ms. Anna Rossetti of the Basic Skills Unit, NEC, for their most helpful comments. May I also express my gratitude to colleagues at Carlton le Willows School and in particular to my wife and family for bearing with me during the writing of this book.

Acknowledgements
The map extract on page 78 is reproduced by kind permission of The Ordnance Survey. Crown Copyright Reserved.

The graphs on pages 84 and 85 are adapted from diagrams which appeared in *Which?* magazine, January and June 1978. Reproduced by kind permission of The Consumers' Association.

About this book

Welcome!
We hope you will enjoy this book.

Who is it for?
For anyone who wants to brush up simple maths ideas and discover how to use them. . . at work, at home. . . anywhere.

This book is full of ways to help you decide what to do, just by using a little maths. It is written for the Yorkshire Television series *Numbers at Work,* but you can use it on its own quite easily.

How the book works
On page 5 are details of every chapter. Most of each chapter is built around practical problems, and the maths is used to help solve them. See how the same maths ideas come up in very different places. Then try solving some of *your* problems with maths.

We hope you'll want to do all the chapters, but there's no need to. Just pick the parts you need, and use the last two pages of the book to help you to find where to look.

These arrows will tell you what page to look back to , or forward to , to get more help.

How you work
First, it's not a race, so take your time. Get into the habit of writing things down as you go. And *do* the questions! Don't just read them. You'll find answers at the back, so check up as you finish each page.

Do the Tests as well — they'll tell you better than anyone how you're doing. And check — all the time — right answers are important.

What you need
Some paper — or better still a notebook. A pen or pencil. Some peace and quiet. A simple calculator with instructions will help when you see this sign, ⬡ . And — if you can — a friend to help you when you get stuck!

Extra help
Maybe it's just too hard — there's not enough space here to go into details on everything. Ask at your library, Adult Education Centre or Adult Literacy Centre about how to get more help.

You may find the book *Make it Count* useful. It's also produced by the National Extension College. Write direct to N.E.C., 18 Brooklands Avenue, Cambridge for details.

4

Contents - what you will find in each chapter_____

Deciding

This chapter shows some ways we can *check* and then DECIDE what to do. Shops *check* to see how much they have sold. Then they know what to order. Factories *check* their stocks of heating oil. Then they can make sure they don't run out. In the supermarket we *check* which items we buy. Then we can estimate how much we must pay. Maths can help us to *check*. And the maths is not too hard!

Don't forget. You will need paper and pencil.
Get in the habit of writing things down as you work them out — then use the answers at the back of the book as you go.

Making a stock check
Susan Evans is filling in her weekly stock sheets

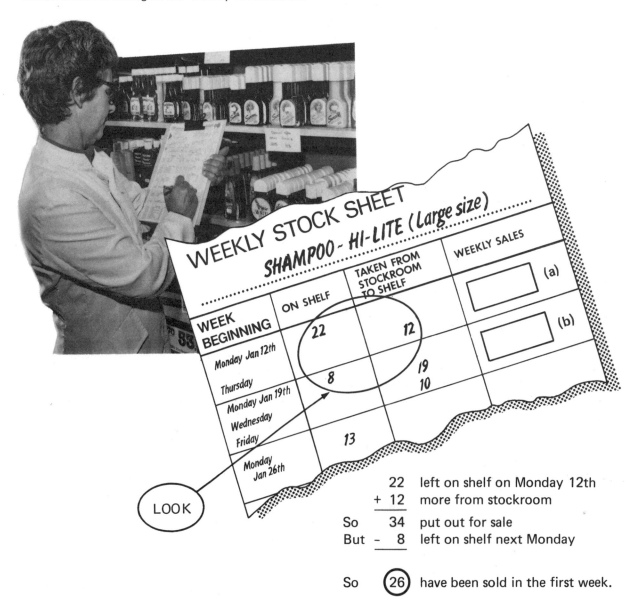

WEEKLY STOCK SHEET
SHAMPOO - HI-LITE (Large size)

WEEK BEGINNING	ON SHELF	TAKEN FROM STOCKROOM TO SHELF	WEEKLY SALES	
				(a)
Monday Jan 12th	22	12		(b)
Thursday	8	19		
Monday Jan 19th		10		
Wednesday				
Friday	13			
Monday Jan 26th				

LOOK

	22	left on shelf on Monday 12th
	+ 12	more from stockroom
So	34	put out for sale
But	− 8	left on shelf next Monday
So	(26)	have been sold in the first week.

Write 26 in box (a)

Now *do the same* for next week. *Write down* what you find out.

Planning ahead

Check your working

```
    8   on shelf on Monday 19th
  +19 ⎫
  +10 ⎭ more from stockroom
  ───
   37   put out for sale
 – 13   left on shelf next Monday
  ───
  (24)  have been sold
```

Write (24) in box (b)

1 Look at some more of Susan's stock sheet. Then work out what goes in each box.

Monday Jan 26 Thursday Saturday	13	 23 8	☐
Monday Feb 2 Thursday	18	 24	☐
Monday Feb 9 Friday	19	 20	☐
Monday Feb 16	12		

2 Look at the numbers you have put in the boxes. They show weekly SALES.
Are any less than 20?
Are any more than 30?

Susan uses her stock sheets to *check* on sales. They help her to *decide* what to order. But it has to be an *estimate*. She can't tell exactly what the sales will be. So she *makes an estimate* based on what she has found out.

3 Now *you* make estimates for the sales for the other two weeks in February.

This chart may help you.

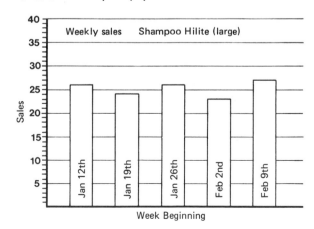

4 This stock list is for a new type of paint. Fill in the boxes

Monday Jan 26 Wednesday	12	 23	☐
Monday Feb 2 Tuesday Friday	6	 25 13	☐
Monday Feb 9 Tuesday Thursday Saturday	8	 14 17 19	☐
Monday Feb 16 Tuesday Thursday Friday	4	 18 22 13	☐
Monday Feb 23 Wednesday Thursday Saturday	9	 16 23 25	☐
Monday Mar 2	12		

5 Look at weekly sales. Are sales staying steady?

6 Make estimates for the sales in the first two weeks in March.

Make a chart like Susan's if it helps.

7 The paint firm delivers supplies every 4 weeks.

How much should Susan ask to be delivered on March 2nd – 100, 200, 300, 400 or 500 tins?

Check answers at the back

Checking on heating oil stocks

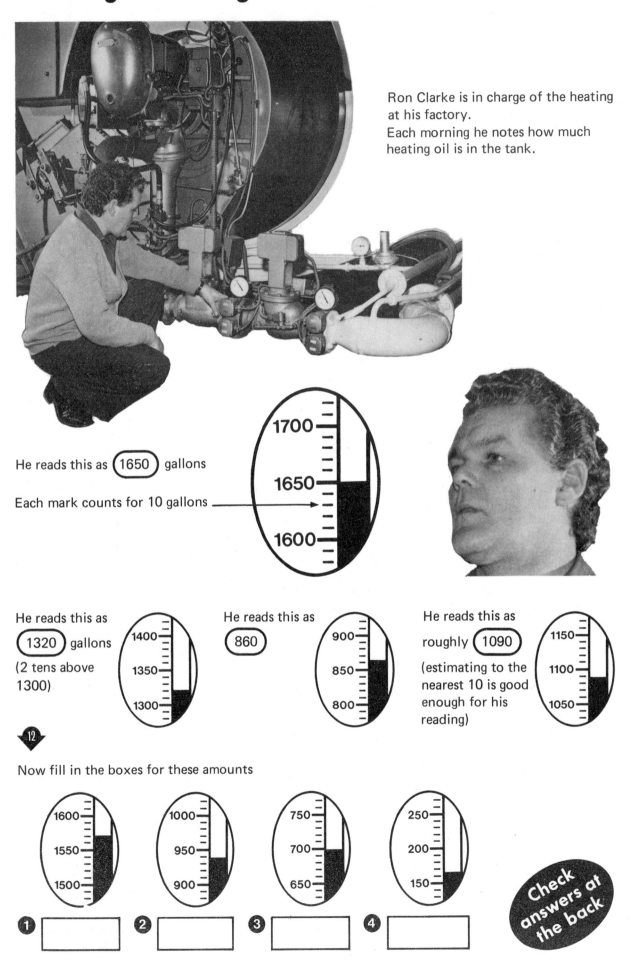

Ron Clarke is in charge of the heating at his factory.
Each morning he notes how much heating oil is in the tank.

He reads this as (1650) gallons

Each mark counts for 10 gallons ⟶

1700
1650
1600

He reads this as
(1320) gallons
(2 tens above 1300)

1400
1350
1300

He reads this as
(860)

900
850
800

He reads this as
roughly (1090)
(estimating to the nearest 10 is good enough for his reading)

1150
1100
1050

Now fill in the boxes for these amounts

1600
1550
1500

1000
950
900

750
700
650

250
200
150

❶

❷

❸

❹

Check answers at the back

Making sure there's enough oil

Ron has to make sure the factory does not run out of oil. So he 'phones the supplier in time.

He must always have (at least 300 gallons) in the tank.

BUT

It takes (2) days for the tanker to come

The tanker delivers (2000) gallons at once

Ron's tank can only hold (2500) gallons.

Here is Ron's record sheet

Date	Reading	Delivery	Amount used
17 Feb	1460		120
18 Feb	1340		130
19 Feb	1210		(a)
20 Feb	1040		(a)
21 Feb	950		(a)
24 Feb	820		(a)
25 Feb	710		(a)
26 Feb	590		(a)
27 Feb	460	2000	(c)
28 Feb	(b)		100
3 Mar	2230		

Ron works out   1460
              − 1340
              ‾‾‾‾‾‾‾
                 120   and fills in 120

5 Fill in the spaces (a) for Ron.

6 2000 gallons arrives on February 27. But he forgets to write down numbers at (b) and (c).
He knows they used 100 gallons on Feb 28. So he works out what goes at (b). And then he fills in space (c). What did he put?

7 Why is no oil used on 22nd and 23rd February?

8 Ron's boss says: "We use roughly 100 gallons each day, or 500 gallons each week. Put in an order once every four weeks for 2000 gallons."
Ron thinks they may run out.
Do you agree?

9 Ron adds up "amount used" each week. Write down what he gets.

10 Use these figures to guess when Ron got the tanker to call again.

11 Ron is off sick, so Ann takes over.
She works out the "weekly amount used" in one step, like this:

17–21 Feb	1460 - 820	= 640
24–28 Feb	820 + 2000 - 2230	=
3–7 Mar	2230 - 1570	=
10–14 Mar	1570 - 980	=
17–21 Mar	980 - 390	=

Fill in the gaps.

12 Ann orders more oil on Monday 23rd March. Is she too late?
Give a reason for your answer.

Check answers at the back

9

Organising a Disco

Multiplying can often save you time. Look it up in *Make it Count* if you're not sure about it — or ask someone for help.

Your calculator may help you, as well. Use it when you see this sign .

Carol Davis is in the Social Club of her factory. The club wants to run a DISCO in aid of club funds.

After a few 'phone calls, Carol has found out

CHURCH HALL
Hire charge £12.00
(max 160 people)

MINERS WELFARE
clubroom
Hire charge £19.50
(max 300 people)

County Junior School
Hire of school hall £18.00
(max 200 people)

BIG SOUND DISCO
£14.00

ACE ROCK BAND AND DISCO
£30.00

The club committee members

> have to *guess* — how many people will come.
> have to *decide* — how much the tickets will be.

Carol says: "Suppose 180 people come, at 40p each".
Then she works out 180 x 40p, like this:

```
  180
x  40
-----
 7200
```

So 180 x 40p = 7200p

or £72.00

She makes a table to help her:

Number of people	Ticket price	Takings
160	40p	(a)
180	40p	£72.00
200	40p	(b)
220	40p	(c)

1 Find what Carol puts in the spaces ⓐ , ⓑ and ⓒ .

Check answers at the back

Planning to make a profit_____

Carol finds the PROFIT, like this

> Profit = Takings — hall and disco costs.

So if takings are £72.00 and the Junior School and Big Sound disco are used,

Hall and disco costs are £18.00 + 14.00 = £32.00

$$\text{and profit} = £72.00 - £32.00$$
$$= £40.00$$

2 Find profit for takings of £80.00, using the Junior School and Big Sound Disco

3 Find profit for takings of £88.00, using the Miners' Welfare and Big Sound Disco

Pete says: "We could charge 60p, if we had the Ace Rock Band, *and* more people would come too."

Keep the points in line

```
£88 . 00
-33 . 50
£54 . 50
```

So he makes a bigger table:

Number of people	Ticket price	Takings (d)	Disco	Hall	Costs (e)	Profit (f)
160	40p	£64.00	Big Sound			
180	40p	£72.00	Big Sound	School	18.00 + 14.00	72.00 – 32.00 = £40.00
200	40p	£80.00	Big Sound			80.00 –
220	40p	£88.00	Big Sound			88.00 – 33.50 = £54.50
200	60p		Ace Rock			
220	60p		Ace Rock			
240	60p		Ace Rock			

He always chooses the cheapest hall big enough for the number of people.

4 Work out the missing amounts in column (d)

5 Now do the same for column (e)

6 Now do the same for column (f)

7 Which would *you* go for, if you were on the Club Committee?

Carol says "What happens if it rains, and people don't come?"

8 Find the profit for only 170 people at 60p, in Miners' Welfare with Ace Rock Band.
Were the Committee right to go for the biggest hall and Rock Band?

9 On the night 227 people came to Miners' Welfare with the Ace Rock Band.
Dave said the profit would be £68.70
Carol thought . . . . "It *must* be more. Pete's table shows me Dave is *bound* to be wrong".
What was the correct profit?

Check answers at the back

11

Multiplying and Rounding off _____

Multiplying

Here are 4 examples to remind you how to do multiplying questions. Make sure you understand them — ask a friend or look it up in *Make it Count* if you're not sure.

Remember | ALWAYS ask yourself, "Does the answer look right?" | even if using a calculator

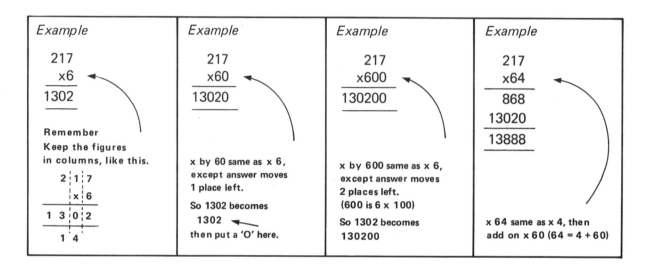

Example	Example	Example	Example
217 x6 ——— 1302 ———	217 x60 ——— 13020 ———	217 x600 ——— 130200 ———	217 x64 ——— 868 13020 ——— 13888 ———
Remember Keep the figures in columns, like this. 2 1 7 x 6 1 3 0 2 1 4	x by 60 same as x 6, except answer moves 1 place left. So 1302 becomes 1302 then put a 'O' here.	x by 600 same as x 6, except answer moves 2 places left. (600 is 6 x 100) So 1302 becomes 130200	x 64 same as x 4, then add on x 60 (64 = 4 + 60)

Now try these. Use a calculator if you like as well.

1 215 x 6   **4** 239 x 40   **7** 239 x 400   **10** 239 x 46

2 239 x 4   **5** 208 x 70   **8** 208 x 700   **11** 517 x 67

3 286 x 10   **6** 286 x 90   **9** 286 x 900   **12** 378 x 96

Often you only need *rough* answers. So *round* off the numbers to make them *simpler*, like this

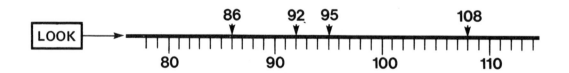

86 is nearer 90 than 80 — so 86 is roughly 90 (rounded to nearest 10)
92 is nearer 90 than 100 — so 92 is roughly 90 (rounded to nearest 10)
108 is nearer 110 than 100 — so 108 is roughly 110 (rounded to nearest 10)
95 is exactly half way between 90 and 100 — it could round *up* to 100, or round *down* to 90.

If in doubt, it's usually better to round up . . . . . so 95 is roughly 100 (to nearest 10).

Round these
 to nearest ten   *Round these*
 to nearest 10p   *Round these*
 to nearest pound

13 27   **17** 431   **21** 32p   **25** £18.10

14 83   **18** 605   **22** 49p   **26** £10.60

15 146   **19** 912   **23** £2.61   **27** £7.29

16 191   **20** 35   **24** £4.77   **28** £9.53

Check answers at the back

Self-Test 1

"I keep a check on what I put in my trolley," says David Scott.

"I round each price to the nearest 10p — then I can keep the figures in my head more easily."

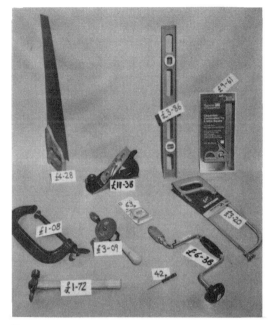

Tool Bar

He has taken 2 loaves (counting as 30p each), 4 packets of butter, 1 packet of Weetabix, 3 packets of tea, 2 packets of sugar, 1 wood saw, 1 hammer and spirit level.

1 Round off each of David's prices to the nearest 10p.

2 Now add these up to get a *rough* total price.

3 He reckons he could buy all the food items in his trolley for £4. Do you agree?

4 Carol wants to buy one of each item on display = 20 altogether. Find the *rough* total price.

5 See how close the rough prices in ② and ④ are by adding the *exact* prices for David and Carol.

6 Now find the price of each item on the Tool Bar, rounding to the nearest pound.

7 Add these up to get a rough total price for these items.

8 Could Carol be certain of buying them for £40?

9 Next week, David has only £3.00 in his pocket. He decides to buy one of as many different food items as he can. How many is this?

10 But he wants coffee, tea, butter, bread, cheese and sugar. How much change will he have after buying these?

Now go over what you have done — then turn to page 94. Give yourself one for each correct answer, except ① and ⑥ (4 for each question).

Chapter 2

Counting at speed

Often, goods are stacked in piles to save space. We need to be able to find out how many sacks are in a stack without counting them one by one. And these stacks are used when loading a lorry as well, to keep the loads secure. Whether it's bags of cement, packets of tea, bottles of beer, hay bales, or bricks, the principle is the same. We can use the same mathematics to count the amounts.

Loading a lorry

Ken Ford is a driver with Hi-Gro fertilizers.
He checks his load before he leaves the yard
The bags are piled on PALLETS
for easy loading.

There are 8 layers of bags on each pallet.
So Ken knows that each pallet holds *8* layers
of *6* bags. And his lorry can carry up to 10 pallets.

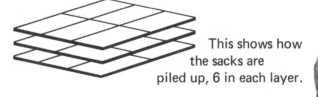

This shows how the sacks are piled up, 6 in each layer.

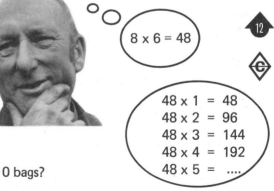

$8 \times 6 = 48$

$48 \times 1 = 48$
$48 \times 2 = 96$
$48 \times 3 = 144$
$48 \times 4 = 192$
$48 \times 5 = \ldots.$

1 How many bags are on each pallet?

2 How many bags can Ken fit on his lorry?

3 How many pallets will he need to give 96 bags?

4 How many pallets will he need to give at least 110 bags?

5 How many pallets will he need to give at least 300 bags?

6 He has to deliver 110 bags to one farmer and 300 to another.
Will 9 pallets be enough?

7 When unloading the 110, 8 bags are damaged.
Has he still enough for the second farmer? How many spares are left?

Check answers at the back

Making up the loads

To save carrying lots of spares, one of the pallets can have less than 48 bags. But there must be complete layers on every pallet, and to allow for damage, some spares have to be included. Ken always has *at least 6 spare bags*. For a delivery of 390 sacks on Monday, Ken must have at least 396 bags

So Ken works out

48 x 8 = 384
48 x 9 = 432

"So I need ☐ full pallets for 384 bags and at least ☐ more bags to make 396.

So I need ☐ more layers of bags".

8 Fill in the boxes for Ken's working. ⟡

While Ken is on the road, his load for Tuesday is made up from these orders:

G. Jones, Fenton, 93 bags	G. H. Daniels, Minton, 173 bags
H. Rider, Clayfield, 71 bags	M. Sugg, Fenton, 88 bags.
S. Holmes, Minton, 84 bags	

9 Can Ken carry enough for all?
If not, which orders should he leave till Wednesday? (Hint: Look at where the orders are going.)

10 What will Ken's load be on Tuesday? (Don't forget the spares!)

11 How many full pallets and extra layers are needed on Tuesday?

12 On Wednesday, Ken takes Rider's order and 400 bags to V. Peters, Wold End.
5 bags are damaged.
How many bags does he still have when he returns?

13 Ken is back in time to make another delivery on Wednesday.
So he is told to deliver 104 bags to a local farmer,
and makes up another 2 full pallets and 2 more layers.
Is Ken *sure* to have enough?

Check answers at the back

Storing hay bales

John and Mary run a mixed farm.
Each year they cut and bale hay from the 20-acre meadow for winter feed.
This summer they make 1520 bales.

They stacked the bales in a box shape
like this, one layer at a time, and put any
left-overs on top.

John says "Lets imagine we make
each layer 18 bales long by
6 bales wide, like this".

1 Work out how many bales there would be in one layer. (Think of it as 6 rows of 18 bales.)

John tries to find how many layers are needed to stack the 1520 bales

"Each layer holds 108, so 10 layers will hold: 108 x 10 = 1080 bales.
So I need more than 10 layers".

2 Work out what 4 more layers would hold.

3 How many layers are needed to hold all the hay?
Mary says "That's too high. You know there should not be more than 10 layers altogether".

4 How many bales must now be in each layer, roughly?

5 Which of these layers will be big enough?
20 x 6, 22 x 6, 25 x 6, 20 x 8, 22 x 8, 25 x 8

6 John makes each layer 22 x 8.
How many complete layers will there be in the stack?

Self-Test 2

Mike Lovatt is in charge of **stock-keeping** at North-East Builders' Supplies.

Each Friday, Mike records stock in the yard. He also notes supplies coming in and out each day.

Roof laths are sold in 12's.
Facing bricks are sold in packs of 720.
Other items are sold in ones.

Each Saturday Mike counts up items in the yard. Then he checks what *is* there against what *should* be there from his stock sheets, like this

N E B S   Stock Sheet

ITEM facing bricks	in units of 720 per pack				
Week ending	In stock ⓐ Monday	In ⓑ	Out ⓒ	In stock ⓓ Saturday	Check ⓐ + ⓑ − ⓒ = ⓓ
27 Sept	0	0	0	0	
4 Oct	0	12	4	8	0 + 12 − 4 = 8
11 Oct	8	15	18	5	
	5				

None in stock on Monday, but 12 packs delivered, then 4 sold so there should be 8 left on Saturday . . . so that's alright

1 Check the next week's figures. Are they alright?

2 A builder wants 70 roof laths. How many bundles does he need?

3 Find how many concrete blocks are in the pile.

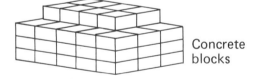

Concrete blocks

plaster

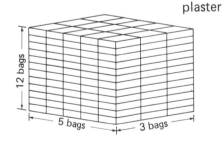

12 bags

5 bags   3 bags

4 Bags of plaster are unloaded. Mike counts 12 layers, 5 bags long and 3 bags wide. How many bags are there?

5 A builder needs enough tiles for the roof of a house as shown. How many tiles does he need? (Each side of the roof needs 16 rows of 32 tiles)

6 Work out the cost of the tiles (5 tiles cost £1.00)

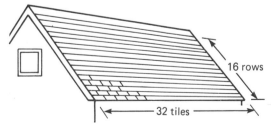

16 rows

32 tiles

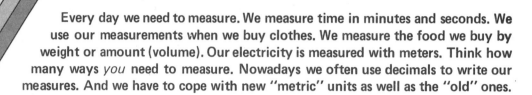

Chapter 3

Every day we need to measure. We measure time in minutes and seconds. We use our measurements when we buy clothes. We measure the food we buy by weight or amount (volume). Our electricity is measured with meters. Think how many ways *you* need to measure. Nowadays we often use decimals to write our measures. And we have to cope with new "metric" units as well as the "old" ones.

Nurse Ann Roberts is reading John's temperature on Monday morning. She reads it as 36.9 degrees (written as 36.9°)

Look at the temperature close up.

The space between 36 and 37 is split into *ten*. Each mark counts for *one-tenth*.

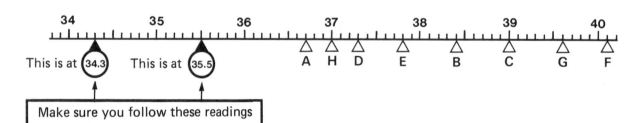

36.9 means   The dot ● is called the | DECIMAL POINT |

Look   36 and 9 tenths

| whole numbers | ● | tenths |

This is at (34.3)   This is at (35.5)   A   H   D   E   B   C   G   F

Make sure you follow these readings

Then fill in the correct figures in these

1 A is at . . . . .

2 B is at . . . . .

3 C is at . . . . .

4 D is at . . . . .

And put in the right letters

5 . . . . . is at 40.1

6 . . . . . is at 37.0

7 . . . . . is at 37.8

Check answers at the back

In hospital

(Look) 38.0 ° means the same as 38 ° (the 'O' means no tenths)

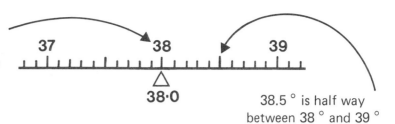

△ 38·0

38.5 ° is half way between 38 ° and 39 °

Give the temperature half-way between

8 34 ° and 35 °

9 41.0 ° and 42.0 °

10 46.0 ° and 47.0 °

11 33 ° and 35 °

12 36 ° and 39 °

13 39 ° and 40 °

Ann keeps a check on John's temperature.
On *Tuesday* morning it is 37.3 °
Ann finds how much it has gone up, like this ——➤
or by saying

"37.3 is 4 tenths more than 36.9"

$$\begin{array}{r} 37\,.\,3\,° \\ -\ 36\,.\,9\,° \\ \hline 0\,.\,4\,° \end{array}$$

Hint: do this like ordinary subtraction but keep the [decimal points] in line

Patient..... *John Davis*		
Ward...... *Hughes (Male)*		
Date	Temp (°C) at 07 00	Change @ daily
4.6.79	36.9	–
5.6.79	37.3	+ 0.4
6.6.79	39.8	
7.6.79	39.3	– 0.5
8.6.79	40.5	+ 1.2
9.6.79	39.0	
10.6.79	38.6	
11.6.79	37.8	
12.6.79	38.2	

14 Work out the changes in temperature. Put '+' for increase. Put '–' for decrease. Write the changes in column ⓐ.

15 Add up all the '+' changes. Take from your answer the total of all the '–' changes. Check that you end up with +1.3. Can you explain why?

16 Ann plots the temperatures on a GRAPH. Part of it is shown below. Look at it carefully, then pencil in the missing points.

17 Keep a record of *your* temperature in ° C. Plot a graph, like Ann's, to show how *your* temperature changes.

(80)

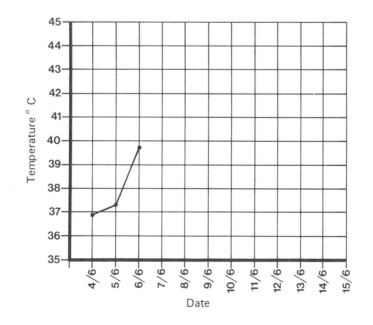

Check answers at the back

Using decimal scales

This is part of a tape measure, showing the piece between 2 m and 3 m marked in tenths

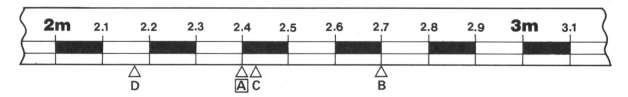

This shows the other side of the same tape, but there are more marks   (Tapes not to scale)

The large marks show *tenths* of a metre   so arrow ☐A☐ is at 2.4 (or 2.40) m

B   is at 2.7 (or 2.70) m

Each *tenth* is divided into *ten* parts again, to give *hundredths* of a metre (small marks)

⎛ Look ⎞

so C is at 2●4┊3 m

and D is at 2●1┊7 m

figures in this column show *tenths* ——————┐
figures in this column show *hundredths* ——————————┐

Now copy and fill in the spaces in these

❶ E is at. . . . .

❷ F is at. . . . .

❸ G is at. . . . .

❹ H is at . . . .

❺ I is at . . . .

❻ J is at. . . . .

❼ 2● . . . . . m is halfway between 2.2 and 2.3.

❽ 2● . . . . . m is halfway between 2.5 and 2.6.

❾ 2● . . . . . m is halfway between 2.3 and 2.4.

The small marks show *centimetres*, so you can also say

C is at 2 m 43 cm (2.43 m)

D is at 2 m 17 cm (2.17 m)

G is at 2 m 80 cm (2.8 or 2.80 m)

H is at 2 m 8 cm   (2.08 m)

Sure you understand? Then read each line below. Say if it is true or not.

❿   2.28 m = 2 m 28 cm

⓫   4.3 m = 4 m 3 cm

⓬   7.2 m is same as 7.20 m

⓭   5.2 m is less than 5.18 m

⓮   0.47 m is same as 47 cm

⓯   61.8 m = 6.18 m

Using decimal scales

Look at part of the tape measure again, shown larger

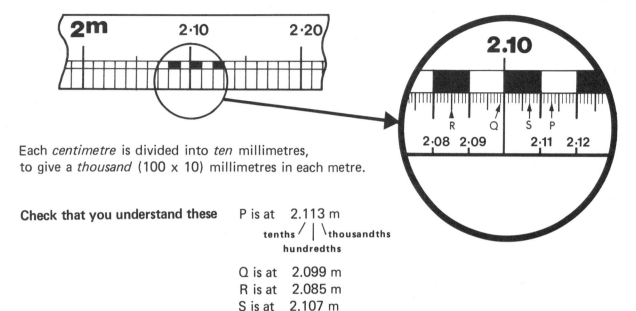

Each *centimetre* is divided into *ten* millimetres,
to give a *thousand* (100 x 10) millimetres in each metre.

Check that you understand these

P is at   2.113 m

  tenths $\diagup$ $|$ $\diagdown$ thousandths
       hundredths

Q is at   2.099 m
R is at   2.085 m
S is at   2.107 m

And these

2.1 m is *same as* 2.10 m (2 m 10 cm) or 2.100 m (2 m 100 mm)

2.107 m is *same as* 2 m 107 mm or 2 m 10 cm 7 mm

2.085 m is *same as* 2 m 85 mm or 2 m 8 cm 5 mm

16 Which is bigger: 2.85 m or 2.085 m?

17 Which is bigger: 2.058 m or 2.06 m?

18 Put these in order, smallest first: 2.03 m, 2.3 m, 2.003 m

19 Do the same with these: 2.64 m, 2.075 m, 2.8 m

20 Pick out a pair of equal measurements from these: (a) 2 m 43 cm; (b) 2 m 43 mm; (c) 2.04 m; (d) 2.043 m; (e) 20 m 43 cm.

21 Pick out 2 equal pairs in these: (a) 6 m 135 mm; (b) 6 m 13 cm; (c) 6.135 m; (d) 6.013 m; (e) 6.130 m; (f) 6 m 135 cm.

(**Look**) how a '5' can often mean 'halfway'

2.085 is halfway between 2.08 and 2.09

3.125 is halfway between 3.120 and 3.130

4.645 is halfway between 4.64 and 4.65

22 What is halfway between 2.06 and 2.07?

23 What is halfway between 3.170 and 3.180?

24 What is halfway between 4 and 4.01?

Check answers at the back

21

Metric units – what they feel like

Most metric units work in THOUSANDS, like the ones you've met already. This section deals with units for measuring *length, volume* and *weight*, and how to work out *areas*.

LENGTH: 1000 *millimetres* (mm) make 1 *metre* (m) — and 1000 m make 1 *kilometre* (km). We also use 100 centimetres (cm) make 1 metre — so 1 cm is 10 mm.

Pete remembers what these lengths "look like"

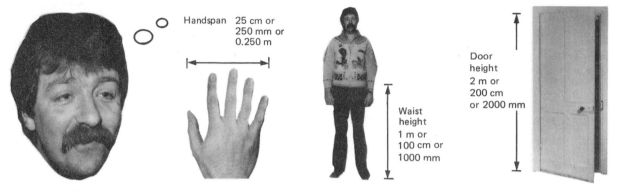

Handspan 25 cm or 250 mm or 0.250 m

Waist height 1 m or 100 cm or 1000 mm

Door height 2 m or 200 cm or 2000 mm

Some of the following measurements just can't be right. Can you spot them?

1. Peter's wife is 1.67 m tall
2. Peter buys a pair of trousers for waist 95 cm
3. Mike measures a brick length as 220 mm
4. Susan's vital statistics are 0.970 m
   0.077 m
   0.098 m

5. Ken's lorry is 375 mm high
6. Ken's foot is 280 mm long
7. Mary buys material in a width of 160 mm
8. Peter's car is 4.5 m long

VOLUME: 1000 millilitres (ml) make 1 litre (l) — and 1000 l make 1 cubic metre (m³)

Susan Jones remembers . . . .

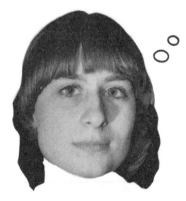

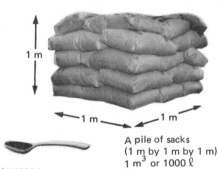

A large tin of oil 5 l (or 5000 ml)

A teacup 200 ml (or 0.2 l)

A teaspoon 5 ml (or 0.005 l)

A pile of sacks (1 m by 1 m by 1 m) 1 m³ or 1000 l

Now do the same with these. Spot the obvious mistakes.

9. A milk bottle holds about 600 ml
10. A large saucepan holds about 2.5 l
11. A large tin of paint holds 5000 ml
12. A coffee mug holds 0.300 l
13. A car petrol tank holds 700 ml

14. A watering can holds about 8 ml
15. An egg cup holds about 0.030 l

Check answers at the back

22

Metric units

WEIGHT

1000 grams (g) make 1 kilogram (kg) — and 1000 kg make 1 tonne (t)

John and Mary remember

A sack of potatoes 25 kg

A bag of sugar 1 kg (1000 g)

A ball of wool 25 g

so
$$1 \text{ g} = 0.001 \text{ kg}$$
and $1 \text{ kg} = 0.001 \text{ t}$

Which of these *must* be wrong?

16 A loaf weighs about 0.8 kg

17 Mary weighs about 60 kg

18 A bag of cement weighs 50 kg (about 0.50 t)

19 An apple weighs about 0.90 kg

20 A brick weighs about 4000 g

21 A pen weighs about 30 g

AREA

We need to measure *areas* to see how much *space* we have.

Area is measured in square metres (m^2), or hectares (10 000 m^2)

m^2 means 'metres squared'

We can find areas by counting the number of square metres which fit in the space.
We add the squares . . . . . or save time by multiplying.

Area 1 m^2

Area 6 m^2 (3 x 2 = 6)

Area 5 m^2 (2.5 x 2 = 5)

(Look)

Wall area 21 m^2 (7 x 3 = 21)

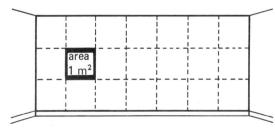

Lawn area *about* 70 m^2 — (10 x 8 = 80, then take off about 10 for the cut off corner)

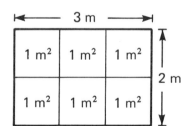

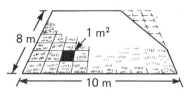

Check answers at the back

22 Now find the area of *your* sitting room walls or *your* lawn.

Another look at taking readings

Often we read off scales which are not marked in tens, but fives or twos.
We read them *as if* they are marked in *tens*.
We *estimate* where the missing other marks are.

Dave weighs in another lorry

He writes 4 t   124 kg = 4.124 t

Write down the weights shown on these scales *in both ways* like Dave.

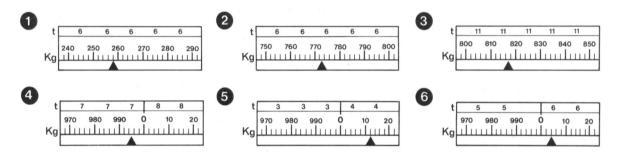

Ann measures temperatures in degrees F

97   98   99   100 °F 101

She reads this as 99.2 ° F   **18**

Write down these temperatures, like Ann.

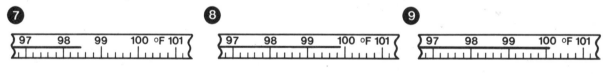

Ken checks the weight of fertilizer bags.
They should all be over 25 kg.

He reads this as 27.3 kg

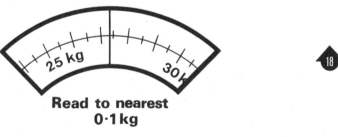

**Read to nearest
0·1 kg**

18

Write down these weights, like Ken.

Check answers at the back

Self-Test 3

Look at each of these scales
Then write down the position of the arrows

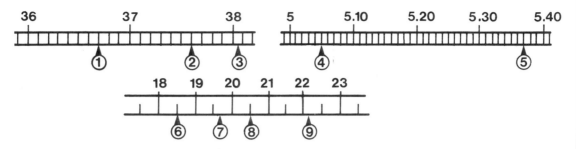

Now see what you can remember about metric units
Fill in the missing words or numbers:

10 A door is about 2. . . . . high

11 There are 1000 . . . . in a metre

12 A teaspoon holds about 5 . . . .

13 2.06 m = 2 metres and 6 . . . .

14 One kilogram = ... grams

15 2.5 m = 2 metres and 500 . . . .

16 mℓ is short for. . . . .

17 24 square metres is written 24 . . . . in short

18 Finally, say which of these *must* be false and which *might* be true.
A tea cup holds 320 ℓ
A lorry can carry up to 4.5 kg
A man is 1.685 m high

A metric summary

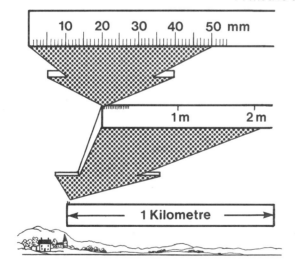

LENGTH	VOLUME	WEIGHT	AREA
1000 millimetres (mm) = 1 metre (m)	1000 millilitres (mℓ) = 1 litre (ℓ)	1000 gram (g) = 1 kilogram (kg)	10 000 square metres (m²) = 1 hectare (ha)
1000 metres (m) = 1 kilometre (km)	1000 litres (ℓ) = 1 cubic metre (m³)	1000 kilograms (kg) = 1 tonne (t)	

Other common units	
10 mm = 1 centimetre (cm)	1 cubic centimetre (cc) = 1 millilitre
100 centimetre (cm) = 1 metre (m)	

A useful conversion
1 mℓ of water weighs 1 g
so
1 ℓ of water weighs 1 kg

This shows how small a millimetre is compared to a kilometre!

Remember   MILLI — means thousandth
                CENTI — means hundredth
                KILO — means a THOUSAND TIMES

Check answers at the back

25

Chapter 4

Working with metric units

There's just no way of escaping metric units!
But there is still the problem of converting from "old" units to the new ones
. . . . In this chapter, we use our knowledge of metric units to sort out some
problems at home and at work . . . . and we get used to making rough conversions.

Using kilograms and tonnes

At North East Builders' Supplies, Mike keeps a check on cement stocks.

Each bag should weigh at least 50 kg
So 2 bags make 100 kg
and 20 bags make 1000 kg or 1 tonne (1 t)

Working with kg and t is just like working in mm and m. There are 1000 kg in 1 tonne, just as 1000 mm = 1 m

This pile has 12 bags so they weigh about 12 x 50 = 600 kg or 0.600 t

Mike practises working in kg and tonnes

50 kg = 0.050 t (or 0.05 t)
100 kg = 0.100 t (or 0.10 t)
150 kg = 0.150 t (or 0.15 t)

Now try these!

Change these into tonnes

1. 750 kg
2. 900 kg
3. 50 kg
4. 1250 kg
5. 75 kg

Change these into kilograms

6. 0.070 t
7. 1.450 t
8. 1.75 t
9. 2.1 t
10. 4.65 t

Find the weights of cement in kg *and* in tonnes

11. 4 bags
12. 9 bags
13. 15 bags
14. 25 bags
15. 34 bags

He also has to measure the lengths of timber roof laths.

16. Find the lengths in metres of these: 5 m 250 mm, 5 m 37 mm, 5 m 41 mm.

17. Find the total length of the three pieces, in metres.

18. Find the lengths remaining after three pieces each 4.775 m long are cut from them.

Check answers at the back

Metric weights

A delivery of cement arrives. Dave checks the lorry in on the weighbridge.

He reads ⎣ 9.792 t ⎦ and writes ⎣ 9.792 tonnes. ⎦

Then he checks it out when it has unloaded

He reads ⎣ 4.288 t ⎦

(19) Find the weight of cement delivered.

(20) Mike checks this against the number of bags delivered.
He says "There were 110 bags, and
20 bags should make at least
1 tonne."
Find what the 110 bags should weigh.

(21) Are the bags underweight?

Each week Mike checks the cement delivery.
He makes sure they are not given "short weight" by checking as in (19), (20) and (21).
He checks that column ⓓ is more than column ⓕ :

N.E.B.M.	Delivery check................. Cement — 50 kg				
Date (a)	Lorry wt. in (tonnes) (b)	wt. out (tonnes)(c)	wt. of goods (d)	no. of bags (e)	minimum weight (f)
14 April	9.792	4.288	5.504·	110	5.500

(22) Dave weighs the lorry on 21st April, as follows   wt. in ⎣ 9.845 t ⎦   wt. out ⎣ 4.190 t ⎦
Copy the chart and fill in column ⓓ .

(23) Mike thought there were 125 bags.
What should they weigh?

(24) He checks and finds 121 bags. Put this in column ⓔ and fill in column ⓕ.

(25) Complete the chart for the weeks shown below.
In each case, check that ⓓ is more than ⓕ .

(a)	(b)			
14 Apr	9.792	4.?		
21 April	9.845	4.190	.655 (d)	
28 April	10.673	4.165		130
5 May	11.491	4.251		145
12 May	10.209	4.199		120

Check answers at the back

Conversion of units

On their farm, John and Mary need to order materials for use around the farm.

Their trouble is that they are used to working with "old" measures (like pints, feet, lbs, etc.) and they have to order in "new" metric measures (like metres, kg, litres, etc.)

They use this table of rough conversions to help — ≈ means "about the same as".
Often a *rough estimate* is all you need.

LENGTH		AREA		VOLUME		WEIGHT	
1 inch	≈ 25 mm (2.5 cm)	1 sq foot	≈ 0.09 m² (900 cm²)	1 fluid oz	≈ 28 mℓ	1 ounce (oz)	≈ 28 g
4 inches	≈ 100 mm (10 cm)	1 acre	≈ 4000 m² (0.4 ha)	1 pint	≈ 600 mℓ (0.6 ℓ)	1 pound (lb)	≈ 450 g (0.45 kg)
1 foot	≈ 30 cm (300 mm) or 0.3 m	1 hectare (ha) (10 000 m²)	≈ 2.5 acres	1 gallon	≈ 4.5 ℓ (4500 mℓ)	1 hundredweight (cwt)	≈ 50 kg
1 yard	≈ 90 cm (0.9 m)	1 m² (square metre)	≈ 11 sq ft	1 litre	≈ 1.7 pint	100 g ≈ 3.5 oz	
1 m	≈ 39 inches or 3 ft 3 in.					250 g ≈ 8.8 oz	
25 cm	≈ 10 inches					1 kg (1000 g)	≈ 2.2 lb (2 lb 3 oz)
						1 tonne	≈ 1 ton

John uses a 2-scale tape measure to help as well with length conversions.

This part shows that 1 ft 10 ins ≈ 560 mm

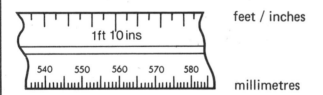

feet / inches

millimetres

They do the conversions like this

| 600 gallons | ≈ | 600 x 4.5 ℓ |
| | = | 2700 ℓ |

| 500 yards | ≈ | 500 x 0.9 m |
| | = | 450 m |

| 16 cwt | ≈ | 16 x 50 kg |
| | = | 800 kg |

7 litres	≈	7 x 1.7 pts
	=	11.9 pts
	≈	12 pts

Now *you* try some. Remember, these figures are only approximate.

1. 700 gal ≈ .... ℓ
2. 8 feet ≈ .... m
3. 20 lbs ≈ .... kg
4. 60 sq ft ≈ .... m²
5. 3 pint ≈ .... ℓ
6. 6 inch ≈ .... mm
7. 2000 sq ft ≈ .... m²

8. 5 cwt ≈ .... kg
9. 15 feet ≈ .... m
10. 10 acres ≈ .... ha
11. 60 g ≈ .... oz
12. 5 fl oz ≈ .... mℓ
13. 4 m ≈ .... ft
14. 10 oz ≈ .... g

Check answers at the back

Conversion of units

Foster farm supplies deliver regularly
John says "....... but this pack is a different size, It's gone metric".

Together Mary and John check the old and new sizes,
to see which pack sizes are larger, which are smaller,
which are about the same.

See if they were right!
Correct any that are wrong

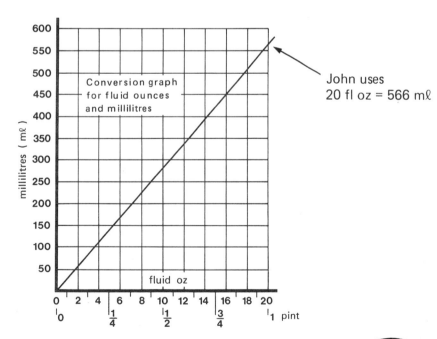

⑮ 5 ℓ is a bit more than 1 gallon

⑯ 10 lb is about the same as 22 kg

⑰ 600 m is more than 600 yards

⑱ 250 kg is about ¼ ton

⑲ ½ pt is less than 500 mℓ

⑳ 60 hectares is less than 120 acres

㉑ 2 m³ is about 70 cu ft

㉒ 400 mℓ is about ⅔ rds of a pint

㉓ 600 g is more than 1 lb

㉔ 6 ft by 4 ft is about 1.8 m x 1.2 m

㉕ Look at the labels on bottles and tins in the shops. Get in the habit of looking at the *amounts* (in grams, mℓ and so on) to get the *feel* of metric units. Sometimes "ccs" are used to measure quantity. Don't worry. 1 cc is same as 1 mℓ.

John uses this conversion graph to change fluid ounces to millilitres.

See is you can see how it works
(page 80 may help).

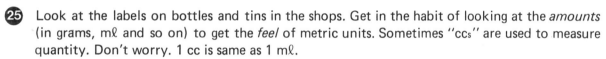

Conversion graph for fluid ounces and millilitres

millilitres (mℓ)

fluid oz

John uses
20 fl oz = 566 mℓ

Exact conversions

It's unlikely you will have to do accurate conversions, except for *lengths*.

Here the 2-scale tape measure is often the best method (see left-hand page).

If you *must* do exact conversions for anything else,
look up the correct figures in the library, technical press or even your diary.

Check answers at the back

Decorating costs

Carol Davis is helping to repaint the walls of their Works Social Centre.
She makes a rough estimate of how much paint is needed
— so she can see how much it will cost.

How many tins will we need?

Special offer £6.59
5 litres white paint
Good cover ! ! ! ! !
1 litre per 17 m^2

She measures the *total length* of the wall in *metres*.

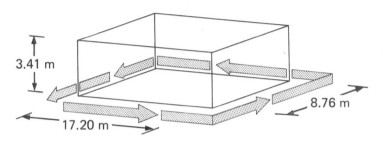

3.41 m

8.76 m

17.20 m

Then she measures the wall height in *metres*.
She multiplies these together, to give the rough wall
area in *m²* (square metres)

Carol writes it down like this

Wall length   =   8.76 + 17.20 + 8.76 + 17.20

              = 51.92 m

              ≈ 52 m

Height   ≈ 3.4 m

Make sure the points are in line

```
    8 . 76
+  17 . 20
+   8 . 76
+  17 . 20
  ─────────
   51 . 92
```

So wall area roughly 52 x 3.4

= 176.8 m²

```
      52
   x  34
   ─────
    1560
     208
   ─────
    1768
```

Carol checks where the decimal point goes in two ways.

Method 1.   52 x 3.4 is roughly 50 x 3 which is 150, so answer is a bit more than 150.

Method 2.   For multiplying   — number of figures to right of decimal point in question
                    = number of figures to right of decimal point in answer

            3 . 4
        176 . 8

Carol knows that 10 17's make 170, so about ☐10 ℓ☐ are needed for one coat or ☐20 ℓ☐ for two coats

So Carol gets 5, 5 ℓ tins just to be on the safe side and cost is 5 x £6.59 = <u>£32.95</u>

Decorating costs

Pete says "Don't forget the entrance hall .

It measures 4.73 m long, 3.24 m wide and 2.88 m high".

1 Find the total wall length for the entrance hall (all 4 walls).

2 Find the rough wall area, like Carol did.

3 Estimate the amount of paint in litres needed for 2 coats.

Pete says "What about using Polypaint?

Each 2½ ℓ tin costs £2.34 *and* it covers over 14 m² per litre."

4 Find a rough cost for painting the entrance hall in Polypaint (2 coats)

5 Which looks better value — Polypaint or Duracolor?

6 They decide to sand and seal the floor of the main room (*not* the entrance)
Find the rough floor area (length x width of room)

7 Find the cost from these details

> Sander — including belts — £20 hire
> Sealer — (3 coats required) — 12 m² per litre
>
> Large size   5 ℓ — £7.20
> Small size   1 ℓ — £1.80

8 Carol says "Before we do the floor, let's do the ceilings — both of them."
Pete says "The area is roughly 8 x 17 + 4 x 3 m²"
Carol says "9 x 17 + 5 x 3 would be better." Why?

9 Use Carol's estimate to find the rough area.

10 Now find the rough cost for painting the ceiling with two coats of Polypaint, this time at 12 m² per litre.

11 The committee want to know "Can you do the whole job for less than £200 — floor, walls and ceiling?"
How would you answer?

13 Make an *estimate* for the total cost.

14 Why can't you be more accurate?

Check answers at the back

Spreading fertilizer

Ken Ford is checking the weight of his lorry load of fertilizer bags. Each bag holds 25 kg.

1 How many bags weigh 1 tonne?

2 His new lorry can carry 13 tonnes.

Can he take 480 bags?

3 One order is for 3½ tonnes.
Ken remembers how many bags
hold 1 tonne.
How many bags does Ken
deliver?

Meanwhile John and Mary work out how much fertilizer they will need.

Their large field measures 200 x 400 m

Fertilizer is spread at 25 kg per 1000 m²

200 m

400 m

4 Find the rough area of the field by working out 200 x 400.

5 Find the number of bags needed.

6 Find the weight of fertilizer in tonnes.

7 In the end, they order 100 bags. What will it weigh?

8 They use the extra bags for another field of 60 x 40 m. How many will they need?

9 Will they have enough? (Allow one bag in every ten for spare).

Some farmers order a complete load of fertilizer at once (13 tonnes).

10 Find how many m² a tonne of fertilizer will cover.

11 Convert this area into hectares (ha) (1 ha = 10000 m²)

12 How many hectares would 13 tonnes cover?

13 Roughly how many acres is this?

Check answers at the back

Self-Test 4

John has grown a special grass for cattle feed in his top field. He has 60 cows in the field. He lets them eat down a strip in the field each day, using a moveable electric fence to stop them eating too much grass.

Each cow needs about 40 kg of grass per day. John has found that 1 m² of grass weighs about 3 kg

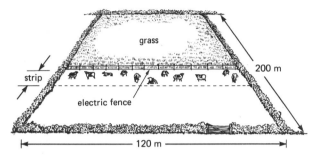

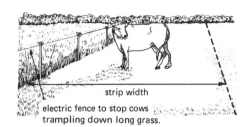

electric fence to stop cows trampling down long grass.

1 How many kg of grass do the cows need each day?

2 How many kg of grass are in a strip 1 m wide?

3 Use your answers to ① and ②
to find the correct strip width.

4 Roughly how many day's feed are in the field?

5 Change these weights to kilograms:
2.315 t, 3.16 t, 4 t (t = tonnes; 1000 kilograms = 1 tonne)

6 Use the table on page 28 to convert each of these measurements
200 ℓ into pints — then into gallons by dividing your answer by 8.
250 g into ounces.

Now turn to page 95 and mark your test (1 for each answer). Did you score 9 or 10?

Now for a mystery!
Here's another conversion graph, but the scales are not labelled!
Who is most likely to use it: Mike, Mary, Ann, Susan, Ken, Pete, Carol or John?
When you're sure, label the scales correctly.

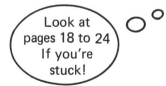

Look at pages 18 to 24 If you're stuck!

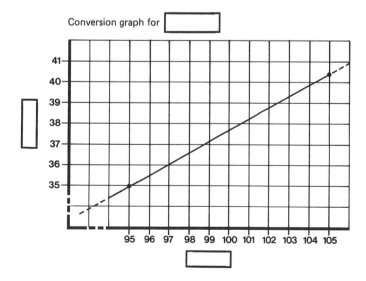

Conversion graph for

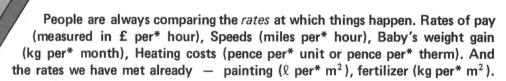

Rks

People are always comparing the *rates* at which things happen. Rates of pay (measured in £ per* hour), Speeds (miles per* hour), Baby's weight gain (kg per* month), Heating costs (pence per* unit or pence per* therm). And the rates we have met already — painting (ℓ per* m²), fertilizer (kg per* m²).

* (Per) means "for every" , *e.g.* £18.00 *per* week means £18.00 *for every* week.

In this section you have to *divide* by numbers like 23, 17, etc. Page 37 will help you. Use a calculator, or ask someone to help you if you have forgotten how to do it. As before a *rough estimate* is often more use than an exact answer.

Susan Jones is coming home on Sunday by car. She reckons she has got *about* 4 gallons in the tank. Her car uses petrol at a rate of about 33 miles per gallon. The distance home is about 110 miles.

Should she try to find a garage or risk not stopping?

When we *make guesses* we have to *allow for errors*

"4" could be 3½ or 4½ gallons
"33" could be 30 or 36 miles

Miles per hour

One summer, Susan is taking the car ferry from Holyhead to Ireland. She has to be at Holyhead at 1:30 pm (13:30 in the timetable)

Living in Nottingham, she has to go 180 miles.

She thinks she can average at least 35 mph for the whole journey.

So the journey time is a bit more than 5 hours
(5 × 35 = 175 miles).

She gives herself 6 hours and starts at 07:30

```
  13:30
− 6:00
  7:30
```

Now check her journey!

1 After 2 hours she has gone 80 miles.
How much further has she to go?
What speed must she average for the rest of the trip?

2 She does another 31 miles in the next hour.
Is she still "ahead of time"?
What speed must she average now?

3 She gets stuck in a traffic jam! The next 11 miles take an hour.
What speed must she average now?

4 The next hour and a half takes her 42 miles further.
How many more miles has she to go?

5 It takes her ½ hour more to reach Holyhead. What was her average for this last section?

6 What was her average speed for the whole journey?
(Total distance ÷ total time in hours)

Meanwhile, Carol is busy planning a coach trip.
She reckons they can't start earlier than 08:30.
They've got to be back by 10 pm (22:00)
(in time for the pubs, her mates tell her).

They have to spend 4 hours wherever they're going to.

They have to stop off for ½ hr on the way and 1 hour on the return.

They can average 35 m.p.h.

Which of these places can they visit? When will they be home?

7 Cambridge (84 miles)

8 L. Windermere (170 miles)

9 Scarborough (120 miles)

10 Bath (140 miles)

Check answers at the back

Filling up car parks

Julie finds the car park full.
It's 3.20 in the afternoon.
All 60 places are taken.
Should she wait — or go to the multistorey (20p)?
She has only ten minutes to get to the bank across the road.

Burton District Council
NORTH STREET CAR PARK

Pay and Display

30 minutes maximum stay

Ticket 5p

She thinks... Some people stay ¼ hr; some chance it and stay more than ½ hour. Let's say, *on average*, each car stays ½ hour.

Some cars will have been there since 2.50 (½ hour ago) .... Others will have arrived only a few minutes ago .....

So, *on average*, a car will arrive once every ½ minute (as 60 half mins make ½ hour)

So, I expect someone to leave once every ½ minute, *on average* — a rate of about *2 cars per minute* .

So she decides to wait. . . . . . and sure enough a car leaves 1 minute later, Julie parks and gets to the bank.

1 South Parade car park holds 120 cars for 40 minutes max.
When it is full what is the average time you might expect to wait for a space?

2 Dave notices . . . . even when there are plenty of spaces, 4 is the most cars which arrive every minute at either car park. . . . . .
Now. . . . . imagine North Street car park is completely empty. What is the *shortest time* it would take to fill up?

3 Now . . . . . . imagine this had just happened . . . . . . How long would you have to wait for a place?

4 Find the shortest time it would take to fill up South Parade car park from empty.

5 How long could you wait for a space if this had just happened?

6 Julie said: "I expect someone to drive away once every ½ minute". Are you surprised that it took her a whole minute to find a space?

In fact, North Street car park was empty at 2.50 that afternoon.
The number of cars arriving in each 5 minutes was:

North Street Car Park (max. 60 cars)		
time	no. of cars arriving	no. of spaces filled
2.50 – 2.55	4	4 **(a)**
2.55 – 3.00	7	11
3.00 – 3.05	11	
3.05 – 3.10	12	
3.10 – 3.15	16	
3.15 – 3.20	10	
3.20 – 3.25		
3.25 – 3.30		4 + 7 = 11

Let's say all cars stay 30 minutes

7 Find the other figures in column (a).

8 How many cars will have left by 3.30?

9 From 3.20 onwards, one car *tries* to park every minute (and goes away if there is no room).
How many spaces will there be at 3.30?

Check answers at the back

Rates in hospital

Nurse Ann Roberts knows how important it is to keep a check on a newborn baby's weight. The baby must put on weight — but not too quickly.

Weight guide		
Age	average wt gain (kg)	average weight gain per week (g)
0–3 months	2.5 kg = 2500 g	200 g
3–6 months	2.0	(a)
6–9 months	1.5	(a)
9–12 months	0.75	(a)

Using the table as a guide:

1 Find the total weight gain in first year.

Ann finds "Average weight gain per week" by dividing by 13 (13 weeks in 3 months)

She says

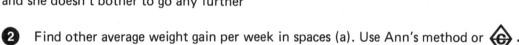

points in a line

13 into 25 is ① and 12 left over

13 into 120 is ⑨ and 3 left over

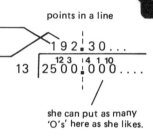

$1\ 9\ 2\ .\ 3\ 0\ ...$

$13\ \overline{)2\ 5\ 0\ 0\ .\ 0\ 0\ 0\ ....}$

13 into 30 is 2 and 4 left over
13 into 40 is 3 and 1 left over
13 into 10 is 0 and 10 left over

and she doesn't bother to go any further

she can put as many 'O's here as she likes.

2 Find other average weight gain per week in spaces (a). Use Ann's method or ◇.

3 Find the average weight gain for a baby from 8 months to 9 months.

⟨81⟩

4 Andrew weighs 4 kg at age 1 month. He weighs 5.5 kg at age 2 months. Is he putting on weight too quickly?

5 Andrew weighs 3 kg at birth. When do you expect him to weigh 9 kg?

6 While milk is its main food, a baby needs roughly 150 ml of milk every day per kg of the baby's weight. How much milk should she/he have at weight 4 kg over a complete week?

Pulse Rates

Ann counts the pulse beats of a patient in 15 seconds (¼ minute). She counts 17 beats.

7 How many "beats per minute" is this?

8 What is 23 beats in 20 secs in "beats per min"?

9 What is 38 beats in 20 secs in "beats per min"?

10 What is 41 beats in 30 secs in "beats per min"?

11 Sarah Green's pulse should not be more than 95. Should Ann get worried if she counts 14 beats in 10 seconds?

12 Perhaps she did not count every beat. Suppose her heart gave 16 beats in 10 secs. Could she be in danger?

Check answers at the back

37

Check up on decimal × and ÷

Try this page if you want to understand what your calculator is up to!

REMEMBER
Common sense is better than any calculator.
Always ask yourself "Does this answer *seem about* right?"

Some quick ways of checking are shown in small type below!
Often a *rough answer* is all you need — you can't *always* be exact . . . . especially when you're trying to work out what *could* happen (how much paint you will need . . . . what your gas bill will be. . . .).

Got the idea? Test yourself by making two lists of calculations — one where the answers are exact, the other where the answers are estimates only.
. . . . . . Now read on . . . . . .
Decimal point — we write it like this 4.7,
But watch out for 4·7 or even 4,7. They all mean the same — 4 whole ones and 7 tenths.

Multiplying ×

Rule. Do multiplying as if there are no decimals; put in the decimal point in the answer as follows:

Examples

$$2.\overset{1}{3} \times 0.\overset{2}{46}$$

$$= 1.\underset{\,}{058}$$

23	
×46	
920	
138	
1058	

Roughly 2 × 0.5 = 1
so answer near 1

Total number of figures to the right of the decimal point before calculating = number of figures to the right of the decimal point in answer.

$$1.\overset{1}{8} \times 400$$

$$= 720.\overset{\,}{0}$$

18	
×400	
7200	

Roughly 2 × 400 = 800
so answer is in hundreds

Dividing ÷

Rule. Dividing is finding *how many* of one number make up another. So if *both* numbers are ×10 (or ×100 or ÷) the answer will stay the same. Use this to change the numbers in the question so that the dividing is as easy as possible.

×10   makes a number move 1 place left
×100 makes a number move 2 places left
÷10   makes a number move 1 place right
÷100 makes a number move 2 places right.

Examples

42 ÷ 0.3 has same answer as

420 ÷ 3 | ×both by 10 |

```
      140
   3 ⌐420
```

= 140

There are about
3 0.3's in 1
so answer will be about
3 × 42 = 126

0.78 ÷ 0.3

= 7.8 ÷ 3 | ×both by 10 |

= 2.6

```
      2.6
   3 ⌐7.8
```

3 × 0.3 = 0.9
so answer a bit
less than 3

Self-Test 5

Susan is looking at the shelves in the shop where she works.
She compares the shampoo prices in two ways.

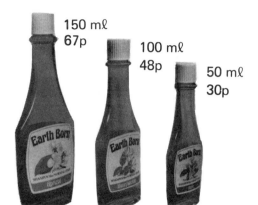

150 mℓ
67p

100 mℓ
48p

50 mℓ
30p

WAY 1

3 small bottles give 150 mℓ costing *90p*

1½ medium bottles give 150 mℓ
costing 48 + 24 = *72p*

1 large bottle gives 150 mℓ costing *67p*

So a *large* bottle gives the best value.

OR WAY 2

Small   1p buys 100 ÷ 48 mℓ — "a bit more than 2 mℓ"   (1.66 .. mℓ

$$\begin{array}{r} \text{or} \quad 1.66.. \\ 3 \overline{)5.00} \quad) \end{array}$$

Medium   1p buys 100 ÷ 48 mℓ — "a bit more than 2 mℓ"   (2.08 mℓ

$$\begin{array}{r} \text{or} \quad 2.08.... \\ 48 \overline{)100.00} \quad) \end{array}$$

Large   1p buys 150 ÷ 67 mℓ — "about 2 and ¼ mℓ"   (2.23... mℓ)
So a bottle of *large* shampoo is best value.

Now see if you can be *sure* to spot the best value. Use either method.

1 The medium and small sizes are "on offer" at and
What size gives the "best value" now?

2 What is the approximate amount she gets for 1 penny in the best value size?

A "new formula" shampoo is brought out in new-style containers. The 'special offer' prices for this shampoo are: Large 140 mℓ at 60p; Medium 90 mℓ at 45p; and Small 45 mℓ at 28p.

3 Find the best value size.

4 What is the approx. number of millilitres per penny for this size?

5 Is it *better value* than the large size before?

6 Most people buy the medium size. What is the approx. no. of mℓ per p?

7 Susan reckons she can get 9 shampoos from a 90 mℓ size.
What is the cost of a shampoo?

8 Susan shampoos her hair twice a week on average.
What is the cost of a year's supply of shampoo for Susan, using 90 mℓ size and assuming the price doesn't change?

9 If she chooses the large (140 mℓ) size instead, find how many shampoos she can get from this size. How long will the bottle last?

10 Find what she would save over the year if she bought the large size instead of the medium.

Percentages

Are you confused about percentages? Do you want to understand how to do percentage calculations? . . . . VAT, pay, fuel saving, even food contents — there is no end to the list of percentage calculations. Just jot down at least ten different situations where you meet percentages. This section shows a "do-it-in-your-head" way to deal with percentages. More exact methods are shown on p. 46.

Customer survey

As part of her work, Susan does a survey of shoppers' views. She questioned 100 people altogether, and noted down their answers.

At the end she added up all the answers from the 100 people and filled in survey sheets. Here is part of one.

Survey sheet   Branch 24

M = male customer
F = female customer

Qu. 8   Do you find the assistants

	All	M	F		All %	M	F
very helpful	12				12		
helpful	62						
unhelpful	18						
very unhelpful?	8						
	100						

Qu. 9   Which of these is the chief reason why you shop here?

	All	M	F		All %	M	F
easy to get to	12						
good range of products	24					24		
low prices	48						
helpful assistants	10						
clean shop	6						
	100						

Susan asked 100 people, so her answers are "out of 100"

% is short for "per cent", which means "out of 100".

So Susan quickly fills in the "All" % columns (some are done already).

1 Copy the sheet and fill in the rest of the "All" % column.

2 Add up the numbers in the All% columns.

3 What is the chief reason why shoppers patronised this shop?

Check answers at the back

Working out approximate percentages_____

Now Susan looks at the replies from men and women.
Were men treated better than women?
Were women more critical?

Out of her 100,
40 were men.

Survey sheet   Branch 24

Qu. 8   Do you find the assistants

	All	M	F	All	M	F
very helpful	12	8	4	12	20	7
helpful	62	26	36	63		
unhelpful	18	6	12	17		
very unhelpful?	8	0	8	8		
	100	40	60			

Qu. 9   Which of these is the chief reason why you shop here?

	All	M	F	All	M	F
easy to get to	12	4	8	12		
good range of products	24	18	6	24		
low prices	48	6	42	48		
helpful assistants	10	8	2	10		
clean shop	6	4	2	6		
	100	40	60			

$$\frac{8}{40} \times 100 = \frac{800}{40} = 20$$

$$\frac{26}{40} \times 100 =$$

Now Susan works out the % figures for men and women.

She says "There were 40 men, so . . . .

8 out of 40 is the same as
16 out of 80 or
24 out of 120 . . . . .

So that is 20 out of 100

or [20%]

next   "26 out of 40 is same as
52 out of 80 or
78 out of 120 . . . . .

that is 65 out of 100

or [65%]

 71

4 Fill in the M% column for question 8 in the survey.

5 Fill in the M% column for question 9 in the survey.

All these figures worked out exactly. Often they don't, but usually it is good enough to round to the nearest whole number.

Susan does the women's answers in the same way.
She says "4 out of 60 is the same as
8 out of 120
or about 7 out of 100, that is 7 per cent"

She works it out roughly by saying
"100 is closer to 120 than 60, so my % will be nearer 8 than 4. I'll say 7".

6 Copy and fill in her working for 36 out of 60

36 out of 60 is same as [ ] out of 120 or about [ ] out of 100,

that is [ ] per cent

Check answers at the back

7 Fill in the rest of the % columns for questions 8 and 9 in Susan's survey.

Finding amounts starting from percentages

"Per cent" means "per hundred".

So 1% means 1 hundredth (divide by 100) — in money "a penny for every pound".

1% of 680 g = 6.8 g   1% of 500 mℓ = 5 mℓ
1% of 120 miles = 1.2 miles
1% of 5800 = 58   1% of £320 = £3.20   1% of £64 = £0.64 or 64p
1% of £17 = £0.17 or 17p   1% of £175.20 = £1.752 or £1.75

Carol remembers . . . .

Some percentages convert into fractions easily. You probably know these:

50% = $\frac{1}{2}$ 25% = $\frac{1}{4}$ 20% = $\frac{1}{5}$ 10% = $\frac{1}{10}$

and, of course, 100% = a whole.

So

> "25% increase in orders" means orders up by $\frac{1}{4}$ of what they were (25 out of 100
> = 1 out of 4 = $\frac{1}{4}$)
>
> "20% drop in sales" means sales down by $\frac{1}{5}$ (one-fifth) of what they were (20 out of 100
> = 1 out of 5 = . $\frac{1}{5}$)
>
> "100% success in sales" means no failure! ($\frac{100}{100}$)
>
> "10% increase in attendance" means attendance figures up by a tenth (10 out of 100 =
> 1 out of 10 = $\frac{1}{10}$)

So as examples

Prices up by 50% (add on a half) means   £4 goes up by £2 to £6
                                           12p goes up by 6p to 18p

Sale 25% off (take off a quarter) means   £4 down by £1 to £3
                                           £6 down by £1.50 to £4.50
                                           £72 down by £18 to £54

What is 6% of £760?
1% of £760 = £7.60, so 6% of £760
(÷ by 100)
= 7.60 × 6
= £45.60

What is 15% of £64?
15% of £64 = 15 × 0.64
= £9.60

"BUT you can *always* work out percentages starting from 1% like this"

Now *you* try some

1 6% of £85   **4** 15% of £72   **7** 50% of £145

2 6% of £670   **5** 15% of £450   **8** 20% of £68

3 15% of £61   **6** 15% of £3.60   **9** 25% of £176

Check answers at the back

Percentage calculations

" ✹ ✹ " says Ken

"I paid £89 last winter. What now?"

1% of £89

Ken works out the *increase* as
$$24 \times £0.89$$
$$= £21.36$$

38

So Ken's bill this winter could be

```
  £89.00
+ £21.36
 £110.36
```

Over £110   " ✹ ✹ " says Ken!

DAILY TIMES

ELECTRICITY PRICE SHOCK! Up by 24%

10 Carol paid £73 last winter. What could she have to pay?

11 Susan only paid £46. What will her new bill be about?

12 Carol says "I could pay about £91 this winter!
I'll save by using less electricity"
She aims to cut the amount she uses by 20%. Find 20% of £91, and give Carol's *reduced* bill.

13 Ken and Susan also aim to save 20% of the electricity they use. Find their *reduced* bill.

14 Rail prices go up by 20%. What is the new price of a ticket which now costs £9.70?

15 A package holiday price goes up from £180 to £234. What is the percentage increase?
(Hint: find 10% of £180.)

Check answers at the back

Using percentages at work

Ron is told by his boss

"Oil consumption must be cut by 5%"

So he keeps a table to check this year's heating consumption against last year's.

Month	Consumption		drop	%drop
	1978	1979		
Jan.	2750	2450	300	11
Feb.	2450			
Mar.	2600			
Apr.	2100			
May	1900			
June	1400			
July				

Ron says

"I can easily check the % drop like this"

1% of 2750 gal = 27.50 gal

and   10% of 2750 gal = 275 gal

so a drop of 300 gal is about 11% of 2750 gal.

42

(A calculator will help you to do this more quickly and accurately. The instruction book will show you how — or ask a friend to help you.)

Here are more figures for 1979

Feb	2350 gal
Mar	2600 gal
Apr	1950 gal
May	1850 gal
June	1500 gal

1 Find the drops for these 5 months.

2 Estimate the % drops, like Ron did (or find them more accurately with a calculator).

3 Copy the chart and fill in the % drops.

Check answers at the back

Percentages at work

The figures for June worry Ron. He hasn't saved any oil. In fact, he's used more!
Ron checks by finding the total oil used Jan — June 1978
and the total oil used Jan — June 1979

4 Work out his figures.

5 Find the total drop in oil consumption.

6 What is 1% of 1978's consumption?

7 What did Ron find as the % drop so far?

8 What is 5% of 13200 gals?

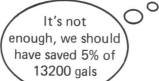
It's not enough, we should have saved 5% of 13200 gals

All over the factory, notices appear

July's figures are

1978	1979
1400 gals	1300 gals

9 Find the 1978 total consumption Jan — July

10 Find the 1979 total consumption Jan — July

11 Find 5% of the 1978 total

12 Fill in the spaces

TARGET	ACTUAL

Are things getting better?

August's figures are

Year	1978	1979
gallons	1600	1450

13 Do the same again as in ⑨ to ⑫.

14 By December 1979, the total oil used was 25410 compared with the 1978 total of 27150.
Find the drop and % drop.
Can Ron have a happy Christmas?

SAVE IT!

DO YOUR BIT TO SAVE ON ENERGY

DON'T WASTE HEAT!
SAVINGS SO FAR
(JAN — JUNE '79)

TARGET	ACTUAL
660	500

Check answers at the back

A closer look at percentages_____

Often we use a calculator to work out problems with percentages. Or we may want more accurate answers anyway.

This section gives the rules for finding percentages exactly. There are quicker methods which some-times work, but these rules *always* work. There are just *two types* of problem.

So read through each section — then try the questions below

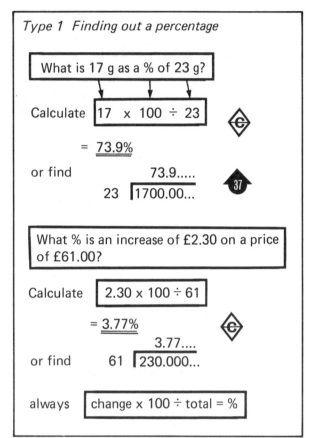

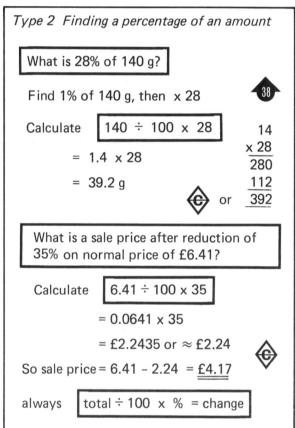

Often you can check % calculations easily, by remembering that 1% is one hundredth, 10% is a tenth, 25% a quarter, 50% a half and 75% three-quarters.

For example

17 is about three-quarters of 23 — so % should be about 75.

One-tenth of £61.00 = £6.10   and   £2.30 is less than half of £6.10,
so £2.30 is less than 5% of £61.00.

28% is roughly a quarter (25%), so 28% of 140 g is about a quarter of it or 35 g.

35% is roughly a third (3 x 35 = 105), so 35% of £6.41 is about a third of it or £2.14.

Now try these

1   What % is 15 g of 60 g?

2   What % is 5 minutes of 1 hour? (change the hour to minutes first)

3   What % is 14p of £2.10 (work in pence)

4   Find 15% of £13.70

Check answers at the back

Shocking percentages

"VAT up from 8% to 15% !" Wait a minute! Just what difference *should* it make?

"15 is nearly double 8. **The price is** nearly doubling".

"No it's only the VAT that's doubling. It's bad, but not that bad".

Who is right?
Susan or Carol?

Carol works it out like this:

"Imagine a radio costing £40 before VAT
The old VAT was 8% of £40

$$= 8 \times 0.40$$

$$= £3.20$$

So price after VAT was £43.20"

"The new VAT is 15% of 40

$$= 15 \times 0.40$$

$$= £6.00$$

So price after VAT is now £46.00"

"So the increase is only £2.80".

Susan agrees with Carol. "I was wrong, but sometimes the % increases seem a lot larger".

Says Carol, "That can be because people don't check the % rises to see if they are right".

Now check these
If they're *wrong*, correct them.

1 Price before 15% VAT = £18. Price after 15% VAT added on = £20.70.

2 Old price £72. New price (due to 12% cost of living increase) £90.

3 Traffic up 20% on last year. Last year 6800 cars per hour. This year 7200 cars per hour.

And try these

4 Butter down 8%. Old price 70p per lb. Find new price.

5 New car prices increase by about 15% *each* year. Cortina costs £3500 now. Find approximate *new* price 2 years from now.

6 Old age pensions are "tied" to the Retail Price Index. At present they are £37.30. If retail prices increase by 10% a year for the next 3 years, find what the pension should be 3 years from now.

Check answers at the back

Slimming

Ken is slimming.
He knows it is important to have a balanced diet, but
He *must* cut down on calories and
He *must* keep protein up.

He says:
"I know beef is supposed to be good but it's expensive . . ."
Let's see how 200 g of beef compares with the other foods.

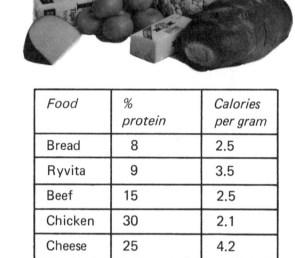

> 100 g of beef gives 15 g of protein.
>
> 200 g of beef gives me 30 g of protein
>
> and 200 x 2.5 = 500 calories. . .
>
> That seems a lot of calories to me."

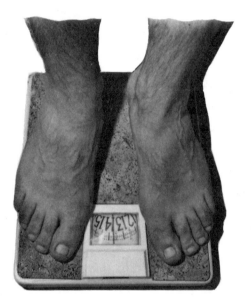

Food	% protein	Calories per gram
Bread	8	2.5
Ryvita	9	3.5
Beef	15	2.5
Chicken	30	2.1
Cheese	25	4.2
Egg	12	1.8
Milk	3	0.7

1 How much chicken will give 30 g of protein?

2 How many calories will this contain?

3 Roughly how much cheese will give 30 g of protein?

4 How many calories will this contain?

5 How much milk will give 30 g of protein?

6 How many calories will this contain?

7 Roughly how many grams of bread will give 30 g of protein.

8 How many calories will this contain?

9 Which looks to be Ken's best bet? Beef, chicken, milk, cheese or bread?

Ken makes up a meal as follows:— 2 slices bread (25 g each), 1 helping of chicken (100 g), a glass of milk (300 g) and 100 g of cheese.

10 Find the total number of calories for the meal.

11 Find how many grams of protein it contains.

12 Using these costs per 100 g, find the cheapest way of obtaining
30 g of protein: Beef 26p, chicken 11p, milk 3p, cheese 17p, bread 4p.

Check answers at the back

Self-Test 6

Try yourself on these % calculations.
Use your calculator if you want to.

1 Find 5% of £12

2 Find 15% of £76

3 Find 8% of £320

4 Find 14% of £6.30

Give these as percentages:

5 3 out of 20

6 13 out of 40

7 6 out of 7

8 1.8 out of 2.5

Find the new prices after these changes

9 Price £12, increase of 5%

10 Price £320, decrease of 8%

11 Price £13.20, increase of 5%

12 Price £45, increase of 15%

And check these answers to see
if they are correct

13 15% VAT on £30 is £4.50

14 15% VAT on £451 is £6.76

15 15% VAT on £2.60 is 39p

16 15% VAT on £573 is £85.95

Finally try these

17 Carol's weight was 140 ℓb. It goes up by 8 ℓb. What is this as a rough %?

18 Oil consumption fell from 1700 gals to 1500 gals. What is the *change* as a % of 1700?

19 "Petrol prices due to increase by 35%." If 4-star costs £1.36 a gallon now, what will it rise to?

20 Only for calculator pundits! Imagine that prices increase by 12% *during each year* for the next 5 years. What would the cost of a new tent be, 5 years from now, if the same design cost £60 now?

Now check your answers. As before, give yourself 1 for each correct answer. Feel pleased if you have 16 or more correct — % calculations are not easy!

Complete this graph.
(*See* pages 44 and 80).
This is called a BLOCK
GRAPH and shows how
much oil is used during
complete months.

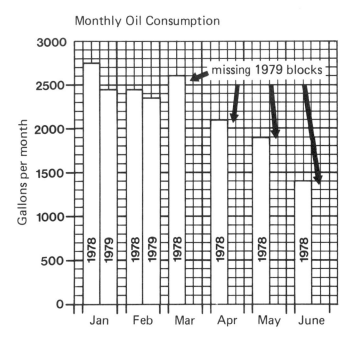

Monthly Oil Consumption

Earning

This chapter is pretty "meaty". Earning money is done in lots of ways — *piecework*, by the *speed* you work; *hourly*, by *how long* you work; and there are overtime, bonuses, tax, etc. to make things more complicated. There is just not enough space to look at any of these in detail. But don't be afraid to ask the "experts" for advice and information. You could gain a lot!

Hourly wages

Ken is paid by the hour.

His basic rate is £1.80 per hour for a 40 hour week.

Any overtime is paid at "time and a half". So each hour that Ken works beyond 40 hours counts as an hour and a half.

So his overtime rate is £1.80 + £0.90 = £2.70 per hour.

Ken checks his pay for week ending 10th April.

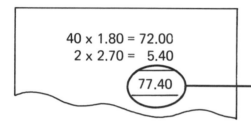

```
40 x 1.80 = 72.00
 2 x 2.70 =  5.40
          ( 77.40 )
```

Employee K. Ford		Basic	O/time	
w.e. 10 April	Hours	40	2	
	Rate £	1.80	2.70	Gross pay
	Pay £	72.00	5.40	77.40

1 Now check these three wage slips for Ken. Are they correct?

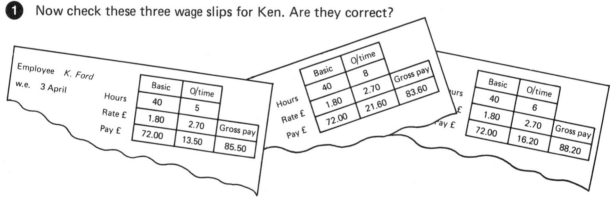

Employee K. Ford		Basic	O/time	
w.e. 3 April	Hours	40	5	
	Rate £	1.80	2.70	Gross pay
	Pay £	72.00	13.50	85.50

	Basic	O/time	
Hours	40	8	Gross pay
Rate £	1.80	2.70	83.60
Pay £	72.00	21.60	

	Basic	O/time	
urs	40	6	
£	1.80	2.70	Gross pay
ay £	72.00	16.20	88.20

Ken's firm takes off tax, insurance etc. before Ken gets his wage packet. Ken's wage packet has his NET PAY (pay after "stoppages")

2 Work out Ken's NET PAY from his complete wage slip.

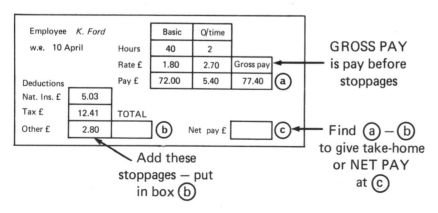

Employee K. Ford		Basic	O/time	
w.e. 10 April	Hours	40	2	
	Rate £	1.80	2.70	Gross pay
Deductions	Pay £	72.00	5.40	77.40 (a)
Nat. Ins. £	5.03			
Tax £	12.41	TOTAL		
Other £	2.80		(b)	Net pay £ (c)

GROSS PAY is pay before stoppages

Add these stoppages — put in box (b)

Find (a) − (b) to give take-home or NET PAY at (c)

Check answers at the back

Clocking in

Ken clocks in at 0730 each morning. and out again each evening.

On Monday his card shows

IN	OUT
0730	1804

This means 4 minutes past 6 — remember it's using the 24 hour clock

Ken thinks. . . . .
From 7:30 to 17:30 is 10 hours and from 17:30 to 18:04 is 34 mins.
So I have done *10 hours 34 minutes*.

(Look) 07:30 to 15:30 is 8 hours.
15:30 to 16:08 is 38 mins.

0730	1608

0730	1621

0730	1600

0730	1657

3 Find the times Ken has worked on the other days.

4 Find the *total* hours worked that week.

5 Find his *overtime* hours.

6 Complete the next payslip for Ken

Employee   *K. Ford*

w.e.   17 April

	Basic	O/time	
Hours	40		
Rate £	1.80	2.70	Gross pay
Pay £			

Deductions

Nat. Ins. £	5.64		
Tax £	14.13	TOTAL	
Other £	2.80		

Net pay £ [ ]

7 At last . . . .
Ken gets a rise.
Complete next week's pay slip.

Employee   *K. Ford*

w.e.   24 April

	Basic	O/time	
Hours	40	5	
Rate £	2.00	3.00	Gross pay
Pay £			

Deductions

Nat. Ins. £	6.17		
Tax £	15.67	TOTAL	
Other £	2.80		

Net pay £ [ ]

Check answers at the back

Bonus payments

Where Carol works, each job is given a "standard time". If Carol can do the job more quickly, she earns a bonus. The times at her factory are worked out with a "Decimal Clock", where each hour is divided into *hundredths* (not sixtieths, like minutes).

So an hour looks like this:

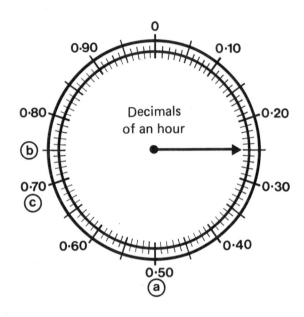

The hand is at 0.25 hours (¼ hr or 15 mins)

ⓐ is at 0.50 hours (½ hr or 30 mins)

ⓑ is at 0.75 hours (¾ hr or 45 mins)

ⓒ is at 0.70 hours ($\frac{7}{10}$ hr or 42 mins)

Standard times

Job A	Job B	Job C	Job D
0.08	0.36	0.17	0.24

Carol does 9 of Job A in 0.65 hrs.
She earns a bonus as 0.65 is less than 9 x 0.08
$$= 0.72 \text{ hrs}$$

Carol does 41 of Job C in 5.30 hrs. She earns a bonus as 5.30 is less than 6.97 hours

```
   17
 x 41
 ----
   17
  680
 ----
  697
```

So 0.17 x 41 = 6 . 97

Now check these times for Carol to see if she earns a bonus.

❶ 6 of Job B in 2.10 hours

❷ 21 of Job C in 3.25 hrs

❸ 18 of Job D in 4.26 hrs

❹ 23 of Job C in 3.71 hrs

Carol is put on Job C full-time.
She wants to earn some bonus, but have a break every 2 hours or so.
She finds how many of Job C she *must* do in each 2 hours.

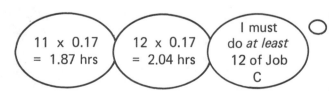

11 x 0.17 = 1.87 hrs   12 x 0.17 = 2.04 hrs   I must do at least 12 of Job C

Bonus payments

(5) Find how much of job B she must do in two hours to earn a bonus.

(6) Find how much of Job A she must do in two hours to earn a bonus.

(7) Find how much of job D she must do in two hours to earn a bonus.

(8) In a 7-hour day, Carol has done 50 of job A and 25 of job C.

Find the standard time for these jobs.

(9) Find how much *bonus* time she has worked, by taking off 7 hours.

Carol finds her *bonus* pay at £2 per *"bonus hour"*, like this

Bonus time	=	4.00 + 4.25 − 7
	=	1.25 hrs
Bonus pay	=	1.25 x 2 = £2.50

(10) Now work out Carol's bonus pay for a week. She works 7 hours each day

Employee...C. Davis..

Date	JOB				Standard time (hr)	Bonus time (hr)	Bonus Pay (£)
	A 0.08 hr	B 0.36 hr	C 0.17 hr	D 0.24 hr			
Monday 7.8 st. time	50 4.00	— —	25 4.25	— —	8.25	1.25	2.50
Tuesday 8.8 st. time	40	—	31	—			
Wed. 9.8 st. time	45	—	28	—			
Thurs. 10.8 st. time	—	12	15	—			
Friday 11.8 st. time	16	10	8	4			

(11) Christmas approaches.
Carol wants to earn at least £20 bonus in a week.
How many *"bonus hours"* must she earn?

(12) By Thursday, she has worked fast enough to earn 8.07 bonus hours.
How many bonus hours must she work on Friday?

(13) Friday is spent on Job D.

Find how many she must complete to obtain her £20 bonus.

Check answers at the back

Looking at tax

"I know tax is 30% now. That means that out of every £10 I earn, I lose £3"

"No, that can't be true, you've forgotten your allowances".

"Look how my tax is worked out," says Ken . . . . .
"I earn about £80 a week after pension payments. . . . . ."

1 Find what this is *per year*. (52 weeks) **12**
"But I get my married man's allowance of £1815 before I start to pay tax". . . .

2 Find out what part of his money Ken has to pay tax on. We call this his "taxable income" — just take £1815 from your answer to **1** . **42**
"And the first £750 is taxed at 25% . . . ."

3 Find 25% of £750.

4 Ken takes £750 from his answer to **2** , and finds 30% of it

What does Ken get?

1% of £1595 is £15.95.
So 30% is £15.95 x 30

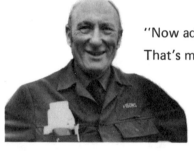

"Now add your answers to **3** and **4**.
That's my tax. Call this **5** ."

5 Write down Ken's tax.
Ken finds what tax he pays *per week* by dividing **5** by 52

$$52 \overline{)666.000} \quad 12.807 \ldots$$

"So I pay £12.81 tax on every £80
That's about 16%."

6 Find what tax Dave pays if he earns £90 per week after pension payments.

7 Why not try this with your own wages?

You may find you're paying even less than you work it out to be.
(p. 55 explains. . . . . .)

Check answers at the back

Looking at tax

Pete says "You can cut your tax more, you know".
"Look at me — I earn £100 per week, but as I have a loan (mortgage) from a building society to buy my house, some of my wage goes on paying off the interest."

"I owe £10,000, and I pay about £100 per month just in interest."
"That's £1200 a year", says Ken.
"Ah . . . but I don't have to pay tax on this, so I have an extra allowance of £1200 before I start to pay tax."

8 Find Pete's taxable income — what part of his money Pete has to pay tax on.

9 Find the tax he pays per year.

10 Find the tax he pays per week.

  As Pete says . . . . "Though I pay £100 a month interest, £30 of this is given back in tax allowances. . . . . . so I only lose £70 a month interest."

11 Now try your hand at working out Dave's tax when his wage goes up to £110 per week and he buys a house with a loan which includes interest of £120 per month.

Ken says: "I also give my parents a bit each week. They've only their pension of £37.30 per week."

12 Find what Ken's parents get each year in pension.

Their tax allowance is £2455. Their pension is less than that — so they don't pay tax.
Ken wants to give them £5 a week extra.
In fact he gives them £3.50 (the remains of £5 after 30% tax), and they *claim the extra* because Ken pays by covenant. So they get £5 after all.
Pete says "That's really good . . . . "

13 Find what Pete has to pay for his parents to get £6.

In fact there are quite a few ways you can cut the tax you pay (legally!). Interest on money you borrow to buy or improve a house is tax free. Making a covenant can help you help someone else — or a charity — as long as the person who is getting help doesn't pay tax.

And some money *you* get (like interest on savings in a Building Society) is tax free. Save-as-you-earn is tax-free as well.
But it's just not possible to give you all the full details here. Ask at your tax office, or get up-to-date details in any of the special Tax guides you can buy at bookshops, or have a look at "Money Which".

Check answers at the back

Working at home

Ken's wife Jean, like all of us, feels she could do with more money! She keeps a good eye on the ads. in the local paper for work she could do at home.

She tells Ken

"I 'phoned up — it's just sewing on labels. I can easily do it.

They said most people can do 50 an hour and at 3p each, that's £1.50 an hour without leaving home".

But when she starts, she can only manage 12 an hour   and she has agreed to do at least 200 a week to keep the job.

1 Find Jean's *actual* pay per hour.

2 Find how many hours she has to work a week.

Ken says: "I know we need the cash — but there must be easier ways — and it's so dull."
"Why not try soft toys — you could always sell them on the market?"

Over a week, Jean buys materials, stuffing and extras costing £3.80.
She makes 5 toys and sells them
all for £2.10 each.
The market stall takes 10% of this.

Ken says: "Don't forget your bus fares — that's 84p altogether"

3 Find what profit Jean made.

4 Find her hourly rate of pay if it took her 7 hours altogether.

Jean decides to carry on!
Next week she uses £4.60 worth of materials to make 6 toys, selling at £2.10 each. It took her 8 hours altogether.

5 Find what profit Jean makes this week (bus fare and market % the same).

6 Find her new hourly rate of pay.

Check answers at the back

Self-Test 7

1 Pete takes a job paying £2.08 per hour for a 40-hour week.

Overtime is paid at "time and a half".

How much does he get for working 46 hours?

2 Over the year his gross pay is £4970. His tax allowances are £1815 plus £205. How much tax does he pay in the year?

3 Sheila is working at a factory now. She is paid on "piecework" — 30p per complete job.

Over a week she completes 236 jobs. What is she paid that week?

4 Sheila's stoppages average about £16 per week, and her gross pay is about £71. She reckons to save 8% of her "take home" pay for holidays.

How long will it take before she has saved £150?

Find out if *your* tax is correct.
Check that *you* understand where each stoppage goes. Your employer is the best person to ask first of all. Then there's your local tax office, or . . .
  your local library, or . . . .
  your local council office, or even
  your local adult literacy centre.

And don't forget. . . .
There are several simply-written booklets on sale just about tax. *Are you sure* you're not paying too much tax?

Chapter 8

Spending at Home

Jean *made* some money by doing extra work at home. Susan *saved* money by driving more carefully. As Carol says, "In the end, what matters is how much you have left after you've paid your bills". It pays to think ahead . . . . . .

Home heating

Carol wants to keep her house snug *and* save on fuel bills. She knows that for central heating, gas or solid fuel are cheaper than oil or electricity.

But she wants to find out what she can *save* with insulation.

She uses these figures showing % heat loss to help her.

Carol says

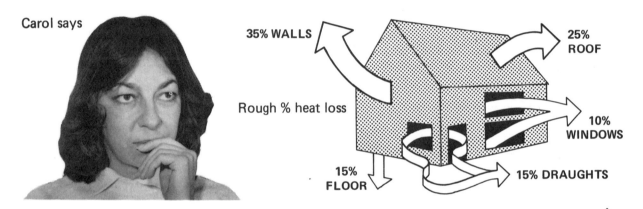

Rough % heat loss

35% WALLS
25% ROOF
10% WINDOWS
15% FLOOR
15% DRAUGHTS

"I pay about £180 per year for heating"
"So I *spend* about 10% of £180 (£18) each year on heat lost through the windows".
"So I can only *save* up to £18 on window insulation".

1. Find *the most* she could ever save by completely insulating the walls, or roof, or just stopping draughts.

2. Actually, she would be lucky to save much more than 60% of these amounts; except for the roof where she could save 75% or more.
Work out possible total savings per year in heating costs.

3. Of course, a lot depends on the *cost* of providing the extra insulation. Work out how many years' savings would be needed to pay for each of these

Cost of extra insulation.
Roof £65; *Walls* £180; *Double glaze windows* £1000; *Draught prevention* £80.

"For the *roof*, I could save 75% of £45

= £33.75

say £34.

So it would take about 2 years to pay for roof insulation out of fuel savings . . . ."

Check answers at the back

Saving on heat

Pete says . . . . . "But you could put money in the Building Society and use the *interest* to help pay the fuel bills

£1000 could easily give you £75 a year interest.
That's more than the cost of the heat going through the window."

(4) **See** how much the other costs would give, if the money was saved in a Building Society instead. Work on interest of 7.5% per year.

(5) What should Carol do first, to cut her heating bill?

Pete reckons he can fit double glazing himself for about £400 for their house.
As he says, "It will cut down noise as well."

So they decide to insulate the roof, put in double glazing, draught proofing and wall insulation.

£65 + £400 + £80 + £180 = £725

We need to borrow about £700

Hire purchase is *easy*

The bank *might* be cheaper

HIRE PURCHASE - Builders' merchant

Repay over 2 years at only £6.50 per month for every £100 borrowed

BANK LOAN

You will have to repay £840, paid back over 2 years, in equal monthly payments.

(6) Which is better value?

Of course, they claim *tax relief* on the £140 interest (as they are using the money for home improvements).
That means they get back 30% of £140 42

(7) What do they pay altogether now? 37

(8) How many years does it take before they save £798 in smaller heating bills?

(9) Have a go on your heating bills. What could you save?

Check answers at the back

Fuel costs

Ron Clarke has just moved house. His new house has gas central heating. He needs to estimate what his gas bill could be. So he finds what gas is used in a week.

Gas is measured in *hundreds of cubic feet*.

cubic feet (hundreds)

Meter reading on 17th October → | 3048 |

Meter reading on 10th October → | 3029 |

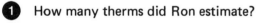

Each hundred cubic feet gives about one THERM of heat

So Ron reckons he has used about
19 therms in the week (3048 − 3029 = 19)

Gas bills are sent out each *quarter* (3 months or 13 weeks).
So he estimates he will use 19 x 13 therms in the quarter October — January.

1 How many therms did Ron estimate?

2 In fact his estimate was on the low side. Why?

Ron noted how much he used each week. It was always between 19 and 40 therms.
"We've used roughly 30 therms each week", he says.

3 Make a better estimate for the number of therms used in the quarter (13 weeks).
Gas prices are roughly
First 50 therms in a quarter — 20p each
All the rest — 16p each

4 Estimate the amount of Ron's bill.

Ron reckons that the bill for the next quarter (Jan — Apr) will be about the same.
He reckons that each of the 2 summer quarters will only need about
half as much gas as each winter quarter.

5 Work out a rough total for all the four bills.

Check answers at the back

60

Gas heating

6 Ron decides to pay his bills by 12 equal monthly payments.

The Gas Board say he should pay £20 per month. Do you agree?

7 What is a fair monthly payment for Ron? (Hint: divide by 12.)

8 Carol's house also has gas for heating and cooking. Being out all day, Carol doesn't use as much gas as Ron (wife and 2 kids!).

Carol also keeps a check on the amount of gas used per week. From October to January the smallest amount used is 12 hundred cubic feet per week and the largest amount used is 28 hundred cubic feet per week.

Estimate the average number of cu ft she used per week.

9 Estimate the number of therms she used for the October to January quarter.

10 What is her gas bill likely to be?

11 She reckons that she will need the same amount of gas for the next quarter but for April-July and July-October, the amounts will only be a third as much.

What are her year's bills likely to total?

12 Estimate a fair monthly payment for Carol.

13 Ron is thinking of putting in an electric shower. He reckons it will save him about 6p each time the shower is used instead of the bath. How many "showers-instead-of-baths" are needed to pay the £120 cost of the shower?

14 Ron's family have 3 showers a day on average. How long will it take to pay for itself?

15 He also installs an extractor fan (cost £25).
How much longer is it before that is paid for through cheaper water heating?

Electricity costs

It's not much use saying that electricity is much more expensive than gas — after all, how *do* you run a hair drier on gas?

There are lots of gadgets and quite ordinary things that can only run on electricity — think of some.

Julie checks on how much things cost to run, like this:

First, she finds the small plate which shows how much *power* they need.

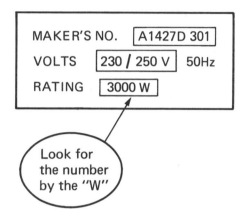

MAKER'S NO.	A1427D 301	
VOLTS	230 / 250 V	50Hz
RATING	3000 W	

Look for the number by the "W"

3000 Watts (W) is same as 3 kilowatt (kW)
(Remember: kilo means x 1000 like kilometre = 1000 metres)

and for each kW, the running cost is about 3p per hour

So a 3 kW kettle would cost about 9p per hour to run, *if heating all the time*.

Here is the *power*. It is 100 Watts

100 W = 0.1 kW (100 W is $\frac{1}{10}$ of 1000 W)

So a 100 W bulb costs 0.3p per hour to run. (3 x 0.1)

 72

38

How long will it run for 10p?

You need to find the number of 0.3p in 10p

= 10 ÷ 0.3

= 100 ÷ 3

≈ 33 hours.

$$\frac{33.3}{3\overline{)100.0}}$$

 38

So it runs for more than a day for just 10p.

Electricity costs

Julie goes round her house and makes a list of electric equipment. For each she finds out its power, in kilowatts (kW) (remember: each kW = 1000 W).

Then she *estimates* the number of hours each is used in a week.
And so she finds how much each will cost her over a week, like this

| cost per hour (p) $\approx$ 3 x power in kW | and then x by number of hours per week

	Power	Cost per hour (a)	Number of hours per week	Cost per week (b)	Cost per quarter (c)
Electric fire	2 kW	6p	30 hours	£1.80	£23.40
Vacuum cleaner	300 W = 0.3 kW		6 hours		
Kettle	3 kW		2 hours		
100 W electric light bulb	100 W = 0.1 kW	0.3p	40 hours		

1 Fill in the gaps in Julie's list in columns (a) and (b) .

2 Now find the cost of each for a quarter (13 weeks).
Put your answers in column (c) .

3 Julie adds up all the lights she has in the house, that she uses every day:

Hall	100 W	20 hours per week
Stairs	100 W	15 hours per week
Bathroom	100 W	12 hours per week
Bedroom	100 W	10 hours per week
Kitchen	100 W	18 hours per week
Sitting Room	100 W	40 hours per week

Find the *total* hours per week for all the lights.

4 Julie finds the cost per week from ③ . What does she obtain?

5 Some items switch themselves on and off.
When the water tank is hot enough, the heater switches itself off until the water cools down again.
Write down two other items which do this.

6 Find how much these cost to run per quarter:

	Possible cost per week	Possible cost per quarter
Chest freezer	31p	
Automatic washing machine	17p	
Colour TV	12p	

Check answers at the back

Do-it-yourself

Pete is making some garden furniture from 75 x 25 mm wood.

1 Make a list of pieces needed to make the table frame (legs, sides and ends).

2 The table is 1.200 m long. Guess how many slats are needed (Pete wants the space between the slats to be between 20 and 30 mm), and correct your guess until the spacing is correct.

3 How long is each slat? (Pete wants the slats 100 mm longer than the table width each side.)

4 Find the *total length* of all the wood needed (work to nearest metre). **21**

5 Find the *total cost* if the wood price is 30p per metre length.

6 He *can* buy the wood in 3 m lengths at 80p each. Would this be cheaper in the end?

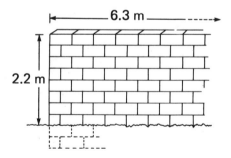

Carol decides to try carpet tiles at home.
Pete wants to compare the cost with carpet.
Carol estimates the number of carpet tiles needed. (Each tile measures 30 cm by 30 cm.)

each tile
30 cm x 30 cm

7 Find what Carol estimated (no. of tiles across room x no. of tiles along). **23**

8 Pete estimates the area for carpet as approx. 5.4 x 3.5 m
What area did he find? Work to the nearest m² (square metre) on the high side.

9 Carpet tiles do *not* need an underlay. Find the cost of tiles at 50p each.

10 The carpet costs £3.65 per m² and the underlay £1.50 per m².
Find the total carpet cost. **38**

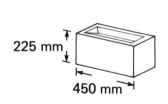

225 mm

450 mm

6.3 m

2.2 m

Dave wants to build a wall
6.3 m by 2.2 m high.
Dave says "I must allow at least
30 cm below ground level".

11 Find the length of 2 blocks and the height of 4 blocks.

12 How many blocks *long* is the wall? **21**

13 How many courses does Dave need?

14 How many blocks does he need (allow 10 extra for "spare").

Check answers at the back

Self-Test 8

1 Ken pays about £250 a year on gas and electricity, split about 70% gas, 30% electricity. Roughly how much is his gas bill?

2 He hopes to save about 15% of his year's gas bill by better insulation. How much would his gas bills add up to then?

3 He is thinking of putting in an 80 W strip light in his kitchen instead of the 150 W bulb. He reckons the light is *on* for about 200 hours a month on average. Find the cost of running the bulb for 200 hours, and the cost of running the strip light for 200 hours. (Look on p. 62 for electricity prices.)

4 Use your answers to ③ to find the rough saving by changing to the strip light, over a month.

5 What is the rough saving per year?

6 Pre-packed coal is for sale at £1.40 per 25 kg bag. Find the cost per tonne.

7 Find the year's coal cost if 1.5 tonne is used altogether.

8 With a more efficient closed-in heater, the fuel cost is £65 per tonne, but only 800 kg are used. Find the new year's coal bill.

9 The new heater cost £80. How long will it take before it has paid for itself?

Another diagram to look at! — often PIE CHARTS help to show percentages compared with each other. (Look back to page 58.)

This is a PIE CHART, to show % heat loss. Put the slices together, and you can see

- Half the heat is lost through floor and walls.
- The roof loss = floor *and* window loss.

You have to imagine that the "pie" is divided into 100 equal slices — so 'walls' have 35 slices (35%), etc.

You see pie charts whenever things are shared out (like amounts spent by your council on police, education, etc.).

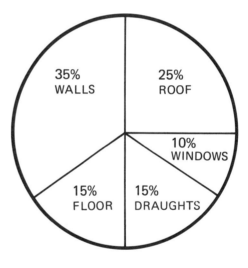

% heat loss from "average" home.

Chapter 9

Getting the right mix

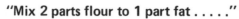

"Mix **2** parts flour to **1** part fat . . . . ."
"Mix **1** measure of red dye to **3** measures of blue dye . . . . ."
"Mix **1** part wheat to **2** parts barley to **1** part oats . . . . ."
"Mix **1** part cement to **3** parts sand . . . . ."
"Mix **1** part oil to **50** parts petrol . . . . ."
All these are recipes for *mixtures* of ingredients.
If the mixture is right, the ingredients are *in the correct proportions,* or we say they
are *in the correct ratio.* **Think of other jobs where the correct mixture is important.**

Pig feed

In making pig feed, John uses this recipe:

1 part by weight wheat to 8 parts by weight barley
to 1 part by weight protein.

The ratio of the mixture is 1 to 8 to 1 by weight

He measures out the correct amount *by weighing.*
1 kg wheat, 8 kg barley, 1 kg protein gives 10 kg.

So one-tenth $(\frac{1}{10})$ is wheat

eight-tenths $(\frac{8}{10})$ is barley

one-tenth $(\frac{1}{10})$ is protein.

Mary says "We need another 50 kg of pig food".
So John finds a tenth of 50 kg.

1   How many kg is this?

There are *8* parts barley to one part wheat. So there will be *8* times as much $(\frac{8}{10})$ barley.

2   How many kg of barley is needed?

3   How many kg of protein is needed?

4   Copy and fill in "$\frac{8}{10}$ is . . . . . times $\frac{1}{10}$."

5   Next week they need 70 kg of feed. So find $\frac{1}{10}$ of 70.

6   How many kg of wheat, barley and protein are needed?

7   Check your answers to (6) by adding them together. What do you get?

Check answers at the back

In the hotel ⸻ In the garden

Mary remembers when she used to cook at the local hotel. Again they worked in (parts by weight.)

To make about 200 g of sweet short pastry:

100 g flour	10 g water
50 g fat	2 g baking powder
25 g egg	1 g colouring
20 g sugar	

8 What does it all add up to?

9 Can you think why it is more than 200 g?

10 The ratio of flour to sugar is 100 to 20 "Oh that's just 5 parts flour to . . . . . . sugar" said Mary. What is missing?

11 For a larger amount of pastry 200 g of flour is needed.

How much sugar is required?

12 400 g of flour would need what weight of eggs?

13 The ratio of egg to sugar is 25 to 20 "That's . . . . . . to 4", says Mary. What is missing?

14 How much egg goes with 60 g of sugar?

15 How much flour would be needed in this mixture in (14)?

16 What weight of pastry would this amount make?

17 If the hotel was down to its last 10 g of baking powder, its last 100 g of eggs and its last 90 g of sugar but had plenty of the other ingredients. . . .
How much pastry could they make?

Outside, John measures out petrol and oil for his 2-stroke motor mower.

The correct mixture is
1 part oil to 50 parts petrol *by volume.*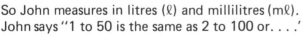

So John measures in litres (ℓ) and millilitres (mℓ).
John says "1 to 50 is the same as 2 to 100 or. . . ."

18 How much oil is needed for 1000 mℓ (1 ℓ) of petrol?
John mixed it in an old 5-litre tin.
He puts in 2.5 ℓ of petrol.

19 How much oil does he need?

20 He uses an old 150 mℓ yogurt carton as a measure.

What fraction of it does he fill with oil?

21 If he filled the carton full of oil, how much petrol would he need?

Check answers at the back

Making concrete

Ken and Dave want to lay a concrete base for a garage at Ken's home. They decide to hire a concrete mixer.

To make the concrete, they use cement, sand, gravel, and water.

They use a cement to sand to gravel mix of 1 to 2 to 4 *by volume*.
They need to find out how much to order.
First they find the volume of concrete they will need.
Ken says "We want it 10 cm thick — that's 0.1 m"

So Dave writes down

    volume = length x breadth x thickness
           = 5 m x 3.5 m x 0.1 m

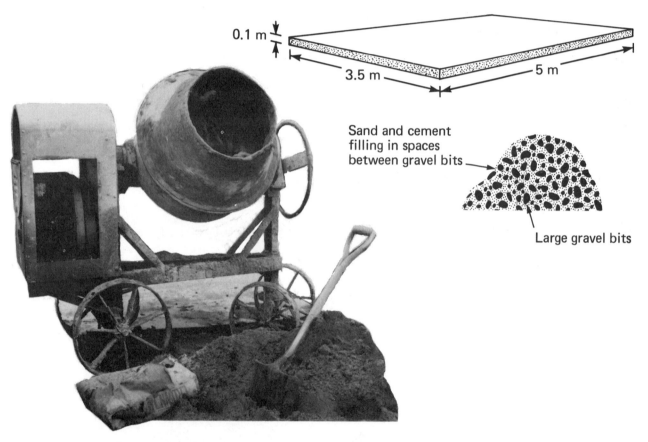

Sand and cement filling in spaces between gravel bits

Large gravel bits

1 What volume does Dave get?

2 Ken says, "Remember the sand and cement just fill the spaces in the gravel, so we need about 1.75 m³ of gravel — say 2 m³."
How many m³ of sand are needed?

3 Cement is measured by weight, not volume. But 1 ℓ of cement weighs 1 kg and there are 1000 ℓ in a m³ . . . . . .
How many kg of cement do they need?

Check answers at the back

4 How many bags of cement will they need (each bag holds 50 kg)?

5 They measure the ingredients by using the same sized shovel.
Ken reckons there are about 10 shovel-fuls of cement in each half bag.
How many shovel-fuls of sand and of gravel are needed to go with this?

6 Ken says "Half a bag of cement holds 25 ℓ."
So how many litres of sand and cement will they need?

Concrete

Hang on Ken the mixer only takes 50 ℓ

"Alright, let's work it out slowly. . . ."
"If I take 1 shovel-ful, that's about 2.5 ℓ", says Ken.
"So 2.5 ℓ of cement need . . . ."

7 How many litres of sand and gravel go with 2.5 ℓ of cement?

Dave says, "Remember the sand and gravel just fill the spaces in between the gravel, so it's the *gravel* volume that's important."

8 How much sand and cement go with 50 ℓ of gravel?

9 Now change that into shovel-fuls for Ken (a shovel-ful = 2.5 ℓ)

10 Dave says, "I knew that ages ago!" Look back to **6** to find how Dave worked it out.

Dave says . . . . "Just a minute — what will all that concrete weigh . . . ? We've got to carry it in barrows"
Ken tells him "1 ℓ of concrete weighs about 2.4 kg".

11 How much is the weight of concrete in the mixer?
How many barrow loads is this? (1 barrow load is no more than 50 kg)

Ken and Dave agree — the calculations are almost as hard work as making the concrete!

So they decide to write down what they worked out — in case they want to do some more concreting sometime.

12 See if you can fill in the gaps
in their table.

Volume of concrete	gravel		sand		cement	
	volume	shovels	volume	shovels	weight	shovels (bags)
50 ℓ (0.050 m^3)	50 ℓ	20	25 ℓ	10	12.5 kg	5 (¼)
100 ℓ (0.100 m^3)	100 ℓ					(½)
500 ℓ (0.5m^3)						
1000 ℓ (1 m^3)						
2 m^3		800	1 m^3		500 kg	(10)

Check answers at the back

Fractions

This is part of a ruler marked in *inches*.

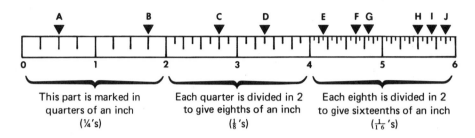

This part is marked in quarters of an inch (¼'s)

Each quarter is divided in 2 to give eighths of an inch (⅛'s)

Each eighth is divided in 2 to give sixteenths of an inch ($\frac{1}{16}$'s)

Instead of dividing into 10, 100, 1000 equal parts like decimals, the inches are divided into 2, 4, 8, 16 equal parts to give $\frac{1}{2}$'s, $\frac{1}{4}$'s, $\frac{1}{8}$'s, $\frac{1}{16}$'s of an inch.

A is at $\frac{2}{4}$, but we call this $\frac{1}{2}$ (it's simpler). We don't need quarters for A.

C is at $2\frac{6}{8}$, but we call this $2\frac{3}{4}$. Here quarters are all we need to fix C.

D is at $3\frac{3}{8}$, and it cannot be simplified.

1   Make a list of the simplest fractions for positions of A up to J.

2   Some of these measures are the same. Match them up
$\frac{3}{8}$, $1\frac{2}{4}$, $1\frac{7}{8}$, $\frac{4}{16}$, $1\frac{1}{2}$, $1\frac{8}{16}$, $\frac{6}{16}$

3   See how each fraction measure is halved on the ruler

$\frac{1}{4}$'s get halved to make $\frac{1}{8}$'s.

$\frac{1}{8}$'s get halved to make $\frac{1}{16}$'s

$\frac{1}{16}$'s get halved to make $\boxed{}$

$\boxed{}$ get halved to make $\frac{1}{64}$'s.

Ken puts aside £60 of his wages each week, like this

Mortgage	Fuel	Housekeeping	Travel	Insurance
£18	£5	£28	£7	£2

He says "£18 out of the £60 goes on the house or $\frac{18}{60}$ as a fraction
That's £9 out of every £30 or $\frac{9}{30}$

or £3 out of every £10 or $\frac{3}{10}$"
He points out "…. So $\frac{18}{60} = \frac{9}{30} = \frac{3}{10}$ — they all mean the same".

4   What fractions does Ken put aside for the other items?

5   Can you make these fractions as *simple* as possible, like Ken did?

6   What fraction of £60 is spent on Fuel and Housekeeping?

7   Make this fraction as simple as you can.

8   What fraction of the £60 is spent on Travel and Insurance?

9   Ken gets a rise, so that he can put aside £70 each week.
He increases the housekeeping to £38.
What fraction is this of £70?

Check answers at the back

Fractions

John and Mary's farm is 60 hectares in area, made up like this:

$\frac{1}{4}$ is winter wheat

$\frac{2}{5}$ are root crops (beet, carrots)

$\frac{1}{5}$ is grassland

$\frac{1}{20}$ is taken up with farmyard, buildings, etc.

the rest is other crops

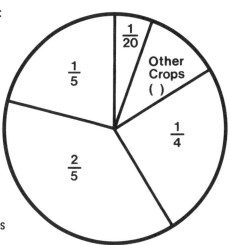

John knows that $\frac{1}{5}$ means a "fifth" (one part out of five)
So he can find how many hectares are grassland by finding $60 \div 5$

$$= 12 \text{ hectares}$$

$$5 \overline{\smash{)}60} \quad 12$$

He says "$\frac{1}{5}$ of 60 is 12 — and 5 12's make 60, of course."
"That means that $\frac{2}{5}$ of 60 is 24" says Mary, "as 2 fifths is double one fifth."
"And $\frac{4}{5}$ would be 48 hectares" says John.

10 Find how many hectares are in winter wheat.

11 Find how many hectares are taken up with the farmyards, buildings, etc.

12 What fraction of the farm is in "other crops"? Give this fraction as simply as you can.

13 They reckon that about $\frac{7}{12}$ of the farm needs proper draining. What area is that?

Remember — You can think of fractions as SHARES — $\frac{2}{5}$ means 2 SHARES out of 5

and *different* fractions can mean the *same* amount

For instance $\frac{2}{5}$ means the same as $\frac{6}{15}$ —
(2 out of 5)   (6 out or 15)

look

$\frac{2}{5}$

2 shares out of 5

$\frac{6}{15}$

6 shares out of 15

Find simpler fractions for these: *Now find these amounts:*

14 $\frac{5}{15}$ **18** $\frac{20}{100}$ **22** $\frac{1}{4}$ of 24   **26** $\frac{1}{7}$ of 63

15 $\frac{3}{12}$ **19** $\frac{75}{100}$ **23** $\frac{3}{4}$ of 24   **27** $\frac{3}{7}$ of 63

16 $\frac{6}{9}$ **20** $\frac{80}{100}$ **24** $\frac{1}{8}$ of 72   **28** $\frac{1}{12}$ of 30

17 $\frac{10}{12}$ **21** $\frac{5}{100}$ **25** $\frac{5}{8}$ of 72   **29** $\frac{5}{12}$ of 30

Check answers at the back

Fraction and decimals _____

Decimals are fractions too, just written differently . . . . Look:

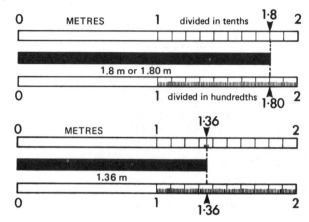

1.8 m   means 1 m and $\frac{8}{10}$ m

each metre is divided into tenths

1.80 m means the same as $\frac{8}{10} = \frac{80}{100}$

1.36 m means 1 m, $\frac{3}{10}$ m and $\frac{6}{100}$ m

   or   1 m, $\frac{30}{100}$ m and $\frac{6}{100}$ m

   or   1 m and $\frac{36}{100}$ m

each metre is divided into hundredths

You can even say "0.95 of my day is spent doing things I hate."

That means $\frac{95}{100}$ or 95% of my day . . . . . . though we hope it's not that bad!

Now you try these:

1   Write 1 and $\frac{4}{10}$ m as a decimal

2   Write 4 and $\frac{23}{100}$ m as a decimal

3   Write 7 and $\frac{6}{100}$ m as a decimal
(Hint: It's *not* 7.6!)

4   Write 2, $\frac{3}{10}$, $\frac{8}{100}$ and $\frac{1}{1000}$ m as a decimal

5   What is 0.75 as a %?

6   What is 0.21 as a %?

7   What is 0.08 as a %?

8   What is 0.4 as a %?

$\frac{1}{8}$ means "1 part out of 8 parts"
So $\frac{1}{8}$ can be made into a decimal by *dividing* — 1 ÷ 8
So $\frac{1}{8}$ = 0.125

$$8\overline{)\begin{array}{l}0.125\\1.000\end{array}}$$

9   Make $\frac{1}{2}$, $\frac{1}{4}$, $\frac{1}{16}$ into decimals in the same way.

    $\frac{3}{8}$ is 3 times $\frac{1}{8}$, so $\frac{3}{8}$ = 3 x 0.125 = 0.375

    or you *can* find it directly by dividing — 3 ÷ 8.

$$8\overline{)\begin{array}{l}0.375\\3.000\end{array}}$$

10   Make $\frac{3}{4}$, $\frac{7}{8}$, $\frac{5}{8}$ and $\frac{3}{16}$ into decimals.

    So you *can* find $\frac{3}{8}$ of £59 by working out 0.375 of £59, which is 0.375 x £59
if you prefer decimals.

11   Use this method to find 0.27 of 61 m.

(same as x)

= £22.155

= £22.15½

or

```
    375
  x  59
-------
   3375
  18750
-------
  22125
```

Check answers at the back

72

Self-Test 9

The gear ratio on a bicycle is (1 to 3) , for top gear. This means (1) turn of the pedals makes the back wheel go round (3) times.

Each turn of the wheel moves the bike 1.7 m.

1 How far does the bike move when the pedals turn once?

2 What speed is the bike going at if the pedals turn once a second? (Give your answer in *metres per second*.)

3 Now convert this speed to *miles per hour* by multiplying by 2.2.

4 In middle gear the ratio is 1 to 2. How many pedal turns per second are needed for the same speed?

5 Mortar for building a brick wall is made from 1 part cement, 1 part lime and 6 parts sand.

A barrowful of mortar needs 32 shovel-fuls.

How many shovels of cement, lime and sand are needed?

6 In the last question, what fraction of this mixture is sand?

7 One of these fractions is the "odd-one-out". Can you find it?

$$\frac{2}{8}\ ,\quad \frac{6}{24}\ ,\quad \frac{4}{10}\ ,\quad \frac{10}{40}\ ,\quad \frac{1}{4}\ ,\quad \frac{5}{20}\ .$$

8 A recipe is given as "Enough for 4 people".
Carol and Pete are having a party. Including themselves, there will be 10 people sitting down to eat.

Change these amounts in the recipe to give correct amounts for 10 people: Flour — 150 g, Fat — 60 g, Milk — 100 ml, Sugar — 20 g.

How fractions, decimals and percentages fit together

These three scales stand for the *same* fractions. Make sure you understand them. The thick lines show tenths.

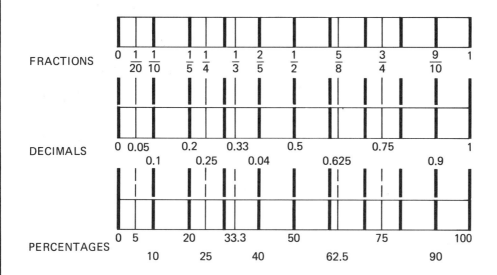

Chapter 10

Maps and plans — you can use them to find out a lot more than you think — if you know the *scale*. And you can use them at home or work — to help planning. Will the new kitchen units fit?
Will the lorry be able to turn into the new loading bay? Just how big *is* that field?
Even in modelling — will the model house look right with those farm animals?

A new kitchen

Carol and Pete are planning their new kitchen. They decide to buy quick assembly kits. The trouble is they're not sure how the furniture will fit — and they don't want to make mistakes.

Carol draws a plan of the kitchen to a scale of 1 to 20.
So 1 centimetre on the plan stands for 20 cm in the kitchen and 5 cm stands for 1 m
(50 mm)   (1000 mm).

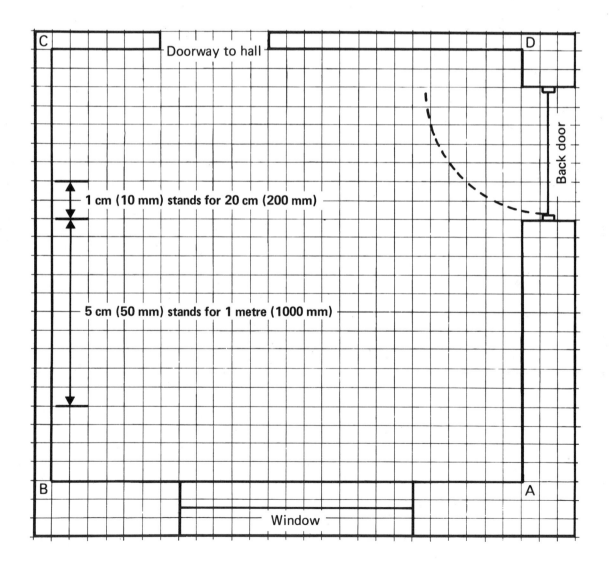

C   — Doorway to hall —   D

Back door

1 cm (10 mm) stands for 20 cm (200 mm)

5 cm (50 mm) stands for 1 metre (1000 mm)

B   A

Window

Fitting up the kitchen

Carol made a table to help in drawing the plan.
She worked out the distances on the plan, like this

Length of kitchen = 2.60 m
                  = 260 cm

So length in plan = 260 ÷ 20

$$\begin{array}{r} 13 \\ 20\overline{)260} \end{array}$$

                  = 13 cm.

Scale 1 to 20

Measurement	Fullsize	On plan	
		cm	No. of squares
Length	2.60 m = 260 cm	13	26
Width			
Window, distance from corner A	0.60 m = 60 cm	(a)	(a)
Window width			
Window distance from corner B	0.70 m = 70 cm	(a)	(a)
Back door distance from corner A			
Back door width			
Hall doorway distance from corner C			
Hall doorway width			

1 Fill in the spaces (a), like Carol did.

2 Carol checks her plan by working backwards, like this. Back door width (on plan) = 3.5 cm (7 squares) so *actual* back door width = 3.5 x 20 = 70 cm or 0.70 m. Do the same as Carol for the other spaces on her table.

Carol and Pete want to fit some of these units. They cut out scale plans of each, like this

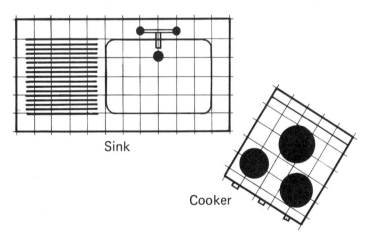

Sink

Cooker

Description	Width (mm)	Depth (mm)
Sink unit	1200	600
Cooker	500	550
Double storage unit with worktop	1200	600
Refrigerator	550	600
Somewhere to have quick meals! (hinged flap or table)	400	600

3 Make scale plans of the other units in their list. (You can trace the squares on the room plan to help — or get some 5 mm squared paper.)

4 Now help them plan their kitchen. Can they fit it all in? What should they do without?

5 Try doing the same for *your* kitchen.

Check answers at the back

Reading a plan

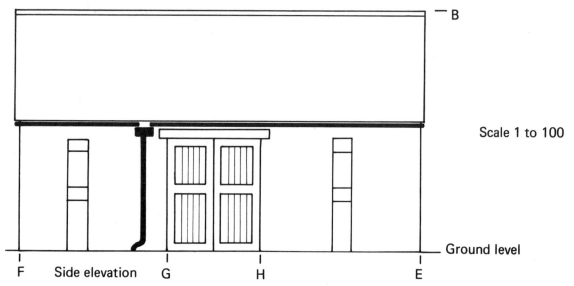

Scale 1 to 100

Ground level

F   Side elevation   G   H   E

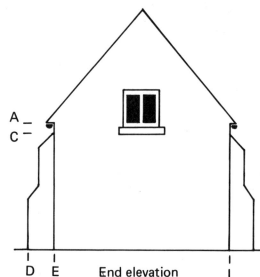

An old barn is to be converted to make a community hall.
The architect draws his side views (elevations) to scale.
You will need a ruler marked in millimetres to help you with this section.

A
C

D E   End elevation   I

1 The scale is 1 to 100 (1:100).
How many millimetres represent 1 metre?

2 Measure these distances in mm. Then convert them to full size measurements.

Height to ridge (B)   Buttress height (C)
Height to eaves (A)   Buttress width (DE)
Door width (GH)   Width (EI)
Length (FE)

3 The walls are about 0.25 m thick. What are the approximate *inside* measurements?

4 Find the approximate inside floor area in m² (remember! length x width)
The builder reckons that the conversion job will cost about £170 per m². They decide to extend the barn to give increased floor area. They want to be able to hold functions for at least 110 people. But the fire officer says each person must have 0.7 m² floor area and they want kitchens etc. of 30 m².

5 What extra area is needed?

6 What is the likely cost?

Check answers at the back

Building an extension

At the fertiliser depot, business is so good that an extension is needed. The trouble is — will the new lorries be able to turn in the tight space?

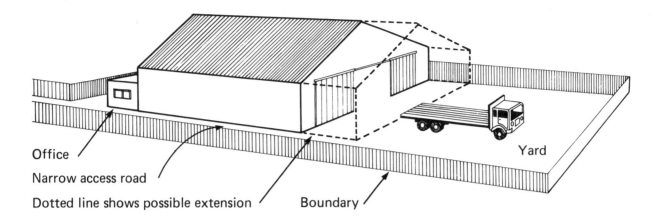

Office

Narrow access road

Dotted line shows possible extension

Boundary

Yard

Ken makes a sketch plan (*not* to scale). It shows how the lorries will have to turn.

Ken finds the distance marked ★ will have to be at least 18 m.

Now Ken draws a plan to a scale of 1 : 400.

He changes each measurement to millimetres, then divides by 400, like this

    5 m = 5000 mm
    on plan 5 m becomes 5000 ÷ 400
                    = 50 ÷ 4
                    = 12.5 mm.

Lorries drive in, then reverse into warehouse

Lorries drive out to A, reverse to B, then drive out of yard past C.

Space for extension

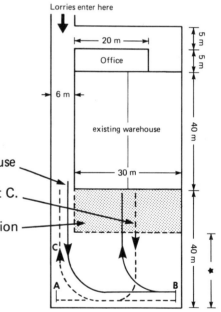

Lorries enter here

Office

existing warehouse

20 m

6 m

5 m

5 m

40 m

30 m

40 m

C

A

B

1 Find the scale plan lengths for the other measurements.

2 On squared paper draw a scale plan of the site.

3 Cut out a piece of card, like this. It represents a lorry, to scale.

6 mm

30 mm

4 Now make your lorry follow the routes in Ken's sketch.

5 Now draw on your plan the extension (shown shaded).

6 Test out the extension with your lorry.
Is there anywhere that two lorries could park, out of the way?

7 Is there room for two lorries to pass in the yard?
Test it out with another card piece.

Check answers at the back

Looking at maps

Often we only need maps to tell us where roads go — to plan a journey, say, or what to see. And we use maps which have distances already filled in, like this

But for more detail, it's best to use Ordnance Survey (OS) maps.

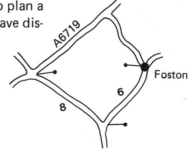

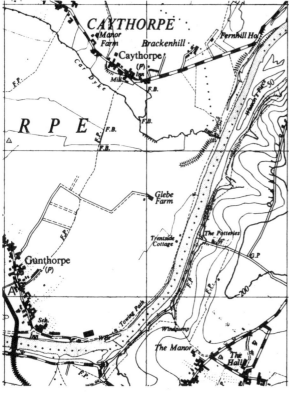

Scale 1 to 25,000   Sheet SK 64

This shows part of Sheet SK64, near Nottingham.

The scale is 1 to 25 000 — good for walkers, as footpaths and hedges are shown.

1 1 cm represents 25 000 cm on the ground. What is this in metres?

2 How many cm stand for 1 kilometre?

3 Use a ruler to find the distance along the footpath from Gunthorpe to Caythorpe over the Car Dyke.

4 Find the length of the footpath circuit running Gunthorpe — Caythorpe — road — back along towing path.

5 How long would you take to walk it? (Assume you can walk at 4 km per hour)

6 How long is the road leading to Glebe Farm?

7 If you left Gunthorpe school at 2.30 pm, when would you expect to reach Trentside Cottage following the circuit of question (4).

8 How wide is the River Trent at this point?

9 The straight lines mark off kilometres. How far apart would they be on a map with a scale of 1 to 50 000?

10 Roughly how far is it from the Footpath end (A) to the School in Gunthorpe?

Now change these map measurements into full-size ones:

11 6 mm, scale 1 to 25 000

12 13 mm, scale 1 to 50 000

13 24 mm, scale 1 to 10 000

And change these full-size measurements to map distances

14 150 m, scale 1 to 10 000

15 40 km, scale 1 to 50 000

16 Change 8 miles to kilometres by multiplying by 1.6, and find the map distance for a scale of 1 to 25 000.

Check answers at the back

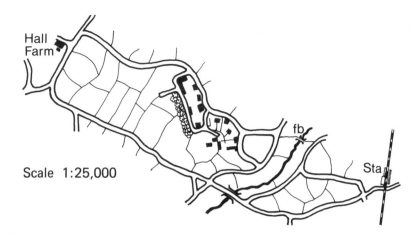

Scale 1:25,000

1 The map scale tells you that 1 cm on the map stands for . . . . . cm on the ground.

2 Convert your answers to ① into metres.

3 Use your answer to find how many centimetres stand for 1 kilometre (1000 m).

4 Measure the distance from Hall Farm to the station along the shortest road, on the map.

5 Now change this distance into kilometres, on the ground. Give your answer to the nearest ½ km.

6 Carol can walk at 4 km per hour. How long should she allow to walk from the Farm to the station?

7 The plan shows part of a notice. Find what scale it is drawn to.

0.625 m

SITE OFFICE

1.5 m

All letters to be 200 mm high.

8 The diagram shows part of the side view of a house extension. Complete the diagram to scale.

9 From your diagram, find the full-size distance from A to B.

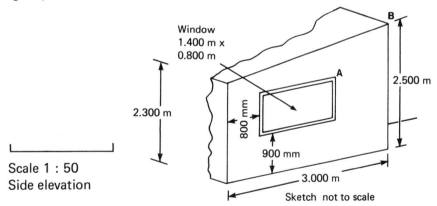

Window
1.400 m x
0.800 m

2.300 m

800 mm

900 mm

Scale 1 : 50
Side elevation

B

2.500 m

3.000 m

Sketch not to scale

Congratulations! You have finished your last test. There is just one more chapter to go . . .
Dont forget to check your test answers! *See* page 94

Chapter 11

Most graphs help to give a clearer picture — but we must be careful to read them clearly. You've already met some earlier in the book. Now you look at them in more detail — and draw some yourself. Learn to read your graphs through and through! It may be that the graph is used to hide more than it reveals! So watch out! You will need some squared paper to work on — 5 mm square or 2 mm square will suit you best.

Graphs we can't argue about — Conversion graphs

Look back to pages 29, 33 and 49 at conversion graphs you've seen already. This page explains how they work and how to make them yourself in more detail.

This graph is simple to make and can be very useful. All you need are 2 points and a straight line. One point is at the 0's. The other stands for 35¼ oz = 1000 g.

A is at 18 oz. Look up to the line and across . . . . and you arrive at 510 g. So 18 oz = 510 g approx.

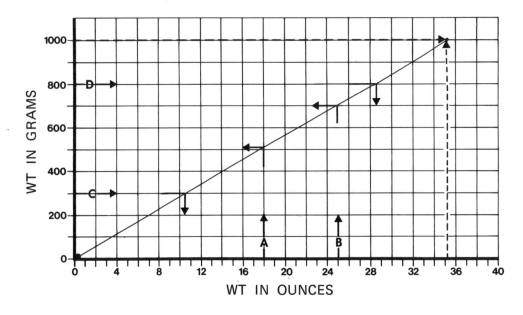

1. Now convert 25 oz to grams (use arrow B)

2. Now convert 300 g and 800 g to ounces (use arrows C and D)

3. On a piece of squared paper, mark on two scales, like the ones above. Label one "inches" and mark in numbers from 0 to 40. Make sure they are spaced out evenly. Label the other "millimetres" and mark in numbers from 0 to 1000 in steps of 100.
Put one dot • in the corner at the 0's. Put the other dot above 39½ ins. and level with 1000 mm.

4. Join up the two dots with a straight line.

5. Use your graph to convert 24 ins, 31 ins, 13½ ins to mm.

6. Now use it to convert 200 mm, 350 mm and 500 mm to inches.

7. Make another conversion graph for pints to litres, using 8.8 pints = 5 litres.

8. Test out your graphs with tins and packets in the kitchen which have 2 measures on them.

Check answers at the back

`Average´graphs

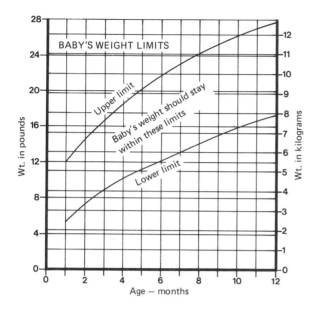

BABY'S WEIGHT LIMITS

Upper limit
Baby's weight should stay within these limits
Lower limit

Wt. in pounds
Wt. in kilograms
Age — months

These graphs aren't exact — they show what is *desirable*.

Babies gain weight at different speeds though their weight *should* stay within these limits.

This graph shows that there isn't a single correct weight for each age — rather a range of weights.

The weights can be read in lbs or kgs. 37

With adults, it's the height that controls your ideal weight.

Although the range *looks* quite wide for each height, the graph has been squeezed up to save paper. (The wiggly line shows this.) In fact there are quite narrow limits for weight (about 6% above and below the ideal weight). 48

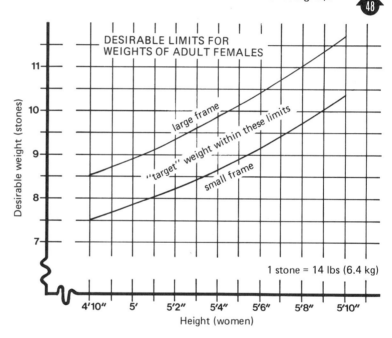

DESIRABLE LIMITS FOR WEIGHTS OF ADULT FEMALES

Desirable weight (stones)

large frame
"target" weight within these limits
small frame

1 stone = 14 lbs (6.4 kg)

4'10"   5'   5'2"   5'4"   5'6"   5'8"   5'10"
Height (women)

Baby weights

1 What are the desirable limits for a baby's weight at 8 months? (Answer in lbs *and* in kg.)

2 A baby weighs 12 lbs. Between what ages would you expect her to be?

3 Use the side scales to find what 12 lbs is in kgs.

4 At nine months, what is the difference between the lower and upper limits of weight? (Answer in kg.)

Adult women's weights

5 A woman of 5'8" weighs 9½ stone. Is she overweight?

6 An "average" weight of 5'3" is 9 stone. What is an "average" weight for 5'7"?

7 Imagine the graph extending to the right. Estimate the lower limit for a woman of 6 ft.

8 A woman of 5'6" weighs 11 stone. Roughly how much weight should she try to lose?

Check answers at the back

Travel diagrams

Two extremes in travelling by train.
This is just a diagram to show the stations on a railway line. You can't work out distances from it. It's *not to scale*. But it's easy to read! Have you seen any other travel maps *not* to scale?

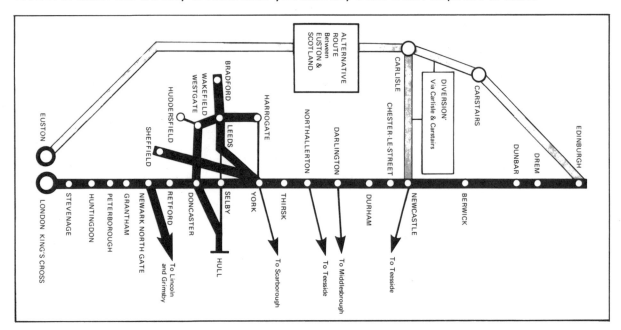

The diagram below is used to plan train timetables. Each sloping line stands for a train.
Full lines — trains on 'fast' lines.
Dotted lines — trains on 'slow' lines.
Lines across — stations.

So Train (b) goes on the slow line to station Q, waits 2 minutes, then carries on to station R.
Meanwhile Train (c) passes it at station Q and reaches S at the same time as (b) reaches R.

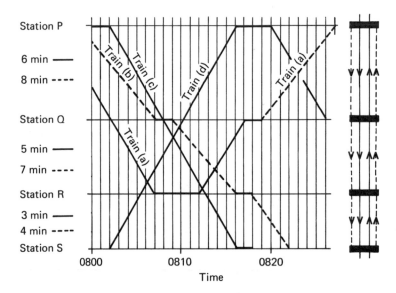

1 How long does train (c) take from station P to station S?

2 How long does (a) wait at R?

3 How long does (a) take from R to P?

4 How long does (b) take from Q to S?

5 At what time does (a) pass (c)?

6 Which trains pass each other twice?

Check answers at the back

Saving money, inflation and borrowing

This graph shows how savings increase

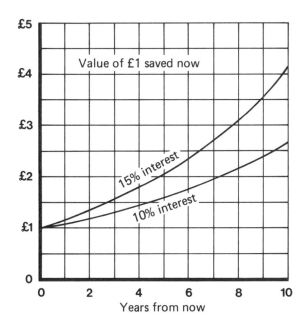

If we save money — in the Post Office, Building Society, Bank deposit account — the extra money we get in return is called INTEREST.

The graph shows that if we put £1 in an account giving 10% interest (that is, the money increases by 10% *each year*), it will have become over £2 in eight years.

The graph curves because each year the interest is more as the 10% is of a larger amount.

So £100 would become about £210 after 8 years!

Good, isn't it?

This graph shows how inflation makes the buying power of money get less

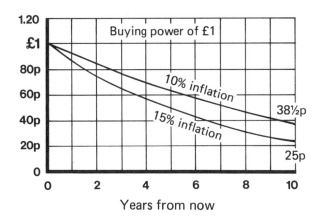

But don't forget, inflation means that money can buy less and less — the value of our money is going down.

Imagine we can save at 10% per year, but inflation runs at 15% per year. . . . What happens after 5 years, say?

Each pound becomes about £1.60, but the buying power of the £1.60 is only about 80p at present values [since buying power is halved in 5 years].

1 How long does it take for £1 to become £2 at 15% interest per year?

2 £200 is saved for 6 years at 10% per year. What does it become?

3 If inflation runs at 10% per year, what is the buying power of £1 after 5 years?

4 The buying power of a pound is halved in 7 years. Estimate the yearly rate of inflation.

When we borrow money — to buy a car, house, furniture — we *pay* interest on the loan. Usually we pay back what we owe (loan + interest) in stages. This means that our loan gets less, year by year, so the interest gets less. It works roughly like this —

We borrow £1000 at 15% interest per year to be paid back in 5 years . . . . .
Our first year's interest will be 15% of £1000, that is £150.
But by next year we will have paid off part of the loan so the interest will be less. . . .

In fact, the total interest we pay over 5 years is about half of £150 x 5, that is about £375. So our yearly repayments will be about $\frac{1}{5}$ of £1375 = 275, or about £23 per month.

How graphs can mislead

Often graphs are correct — but we read too much into them. Look at the graph below.
Is National Panasonic *really* that good?
The % scale makes the graph look true in general. But all this result was based on *just 53* TV sets.
Perhaps the Murphy sample would have done much better if more had been chosen — after all 48 is not many.

Imagine a large warehouse with a stock of thousands of TV sets — half of them going to need repairs within the next 12 months . . . . .
By chance it is quite possible to choose 48 with as little as 18 good or as many as 32 "good" sets.
So the figures could vary from 36% to 64% *by chance*!

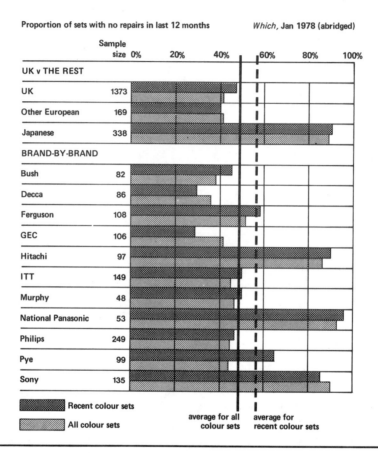

Proportion of sets with no repairs in last 12 months   *Which*, Jan 1978 (abridged)

	Sample size
UK v THE REST	
UK	1373
Other European	169
Japanese	338
BRAND-BY-BRAND	
Bush	82
Decca	86
Ferguson	108
GEC	106
Hitachi	97
ITT	149
Murphy	48
National Panasonic	53
Philips	249
Pye	99
Sony	135

▓ Recent colour sets
▒ All colour sets

average for all colour sets   average for recent colour sets

Here the wine consumption looks high compared with beer. . . . . until you look at the side scales.

Beer increase — 90 pints per head
Wine increase — 9 bottles per head (equivalent to, say, 30 pints of beer).

The side scale for wine is more stretched out than for beer.

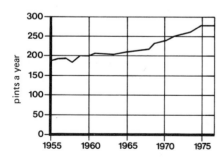

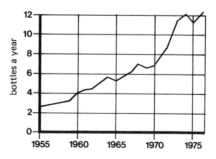

'Which' June 1978

Beer consumption in UK
Consumption per head (excluding home-brewed beer) for people over 15

Wine consumption in UK
Consumption per head (excluding home-made wine) for people over 15. 'Bottle' is 70 cl

Being careful!

Study these two graphs *together*, then read the comments. They show how one graph may only tell a part of a story.

Is our tax on income about the same as in Sweden?

"Yes — you only have to look at the chart — they're both 47%," says Carol.

"Ah, but 47% of what?" asks Julie.

"The Swedish taxes are nearly $\frac{1}{3}$ as much again as ours, overall."(46% compared with 37% in diagram 2.)

Diagram 1: Where the tax comes from (% of total tax collected in 1975 in each country [1])

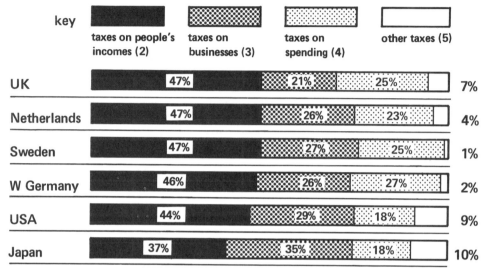

| key | taxes on people's incomes (2) | taxes on businesses (3) | taxes on spending (4) | other taxes (5) |

UK	47%	21%	25%	7%
Netherlands	47%	26%	23%	4%
Sweden	47%	27%	25%	1%
W Germany	46%	26%	27%	2%
USA	44%	29%	18%	9%
Japan	37%	35%	18%	10%

(1) For some countries, percentages don't add up to 100% because of rounding.

(2) Tax on income and capital gains; national insurance contributions paid by employees and self-employed.

(3) Tax on company profits; payroll tax; employers' national insurance contributions; taxes on business property.

(4) VAT: sales tax: customs and excise duties: licence fees.

(5) Including wealth tax, taxes on gifts and inheritances, stamp duty, tax on value of home (eg rates)

Do Japanese businesses pay more tax than ours?

"Yes, look at diagram 1," says Carol — "they pay 35%, we only pay 21%."

"But 35% of what?" asks Julie. "For every £1 worth of goods produced in Japan, 20p goes in tax and 35% of *that* goes in business tax — that is 7p."

"In Britain, 37p goes in tax and 21% of that is business tax — that is 21 x 0.37 = 7.77p ≈ 8p. So *we* pay more in Britain ", says Julie.

DIAGRAM 2: Percentage of GDP paid in taxes and national insurance in 1975

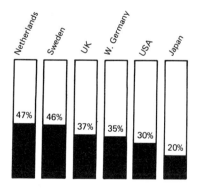

Netherlands	Sweden	UK	W. Germany	USA	Japan
47%	46%	37%	35%	30%	20%

GDP = Gross Domestic Product (total value of goods etc produced each year).

Both graphs from 'Which' June 1978 (abridged).   There is no test for this Chapter

Answers

CHAPTER 1

Page 7

1. 26, 23, 27
2. No, No
3. It seems likely that sales will stay in 20's.
4. 29, 36, 54, 48, 61
5. No, sales are rising
6. A rough estimate would be 70 in the first week and 80 in the second
7. 4 weeks of 70 or 80 could require 300. So order 300, maybe 400 if there's enough storage room.

Pages 8—9

1. 1570
2. 940
3. 700
4. 160
5. 170, 90, 130, 110, 120, 130
6. (b) is 2330. (c) is 130
7. Weekend?
8. Amounts are nearly all more than 100, so they will need *more* than 2000 in 4 weeks. They could run out.
9. 640, 590
10. 17 working days after 27th February gives 23rd March.
11. 590, 660, 590, 590
12. Tanker comes on Wednesday 25th March. Only 110 gallons left. This is *too low*.

Pages 10—11

1. (a) £64.00, (b) £80.00 (c) £88.00
2. £48.00
3. £54.50
4. £120.00, £132.00, £144.00
5. £26.00, £32.00, £33.50, £48.00, £49.50, £49.50
6. £38.00, £48.00, £72.00, £82.50, £94.50
7. -
8. £52.50 — more than profit for 200 at 40p, so this seems the best to go for
9. £86.70

Page 12

1. 1290	6. 25740	11. 34639	16. 190	21. 30p	26. £11
2. 956	7. 95600	12. 36288	17. 430	22. 50p	27. £7
3. 2860	8. 145600	13. 30	18. 610	23. £2.60	28. £10
4. 9560	9. 257400	14. 80	19. 910	24. £4.80	
5. 14560	10. 10994	15. 150	20. 40	25. £18	

CHAPTER 2

Pages 14—15

1. 48
2. 480
3. 2
4. 3
5. 7
6. Yes, he will have 432 bags
7. Yes, he will still have 314 bags. 14 spare bags are left
8. 8, 12, 2
9. No. Leave the Clayfield order. Then he only has to go to 2 villages.
10. 444 bags
11. 9 pallets + 2 extra layers
12. 4 bags
13. Not really. Only 4 spares are included (less than 6).

Page 16

1. 108
2. 432
3. 14 complete layers and some left over bales giving 15 layers altogether
4. 152 or more (152 x 10 = 1520)
5. 20 x 8, 22 x 8, or 25 x 8
6. 8 complete layers + some extra bales.

Answers

CHAPTER 3

Pages 18—19

1. 36.7
2. 38.4
3. 39
4. 37.3
5. F
6. H
7. E
8. 34.5°
9. 41.5°
10. 46.5°
11. 34°
12. 37.5°
13. 39.5°
14. +2.5, −1.5, −0.4, −0.8, + 0.4
15. + changes = 4.5
    − changes = 3.2
    38.2 (last temperature) is 1.3 more than 36.9 (first temperature).

Pages 20—21

1. 2.35 m
2. 2.61 m
3. 2.80 m
4. 2.08 m
5. 2.55 m
6. 2.25 m
7. 2.25 m
8. 2.55 m
9. 2.35 m
10. True
11. False. 4.3 m = 4.30 m, so 4.3 m = 4 m 30 cm
12. True
13. False. 5.2 = 5.20 m, so 5.2 m is *more* than 5.18 m
14. True
15. False. The decimal point is not in the same place
16. 2.85 m
17. 2.06 m
18. 2.003 m, 2.03 m, 2.3 m
19. 2.075 m, 2.64 m, 2.8 m
20. (b) = (d)
21. (b) = (e) ; (a) = (c)
22. 2.065
23. 3.175
24. 4.005

Pages 22—23

1, 2, 3. Alright
4. Bust alright; others far too small
5. No. Should be 3.75 m or 3750 mm
6. Alright
7. No. 1.60 m would be more likely
8, 9, 10, 11, 12. Alright
13. No. 70 ℓ would be more likely
14. No. 8 ℓ would be more likely
15. Alright. 0.030 ℓ = 30 mℓ
16, 17. Alright
18. 50 kg alright but it equals 0.050 t, *not* 0.50 t
19. No
20, 21 Alright
22. Ask someone to check this

Page 24

1. 6 t 258 kg = 6.258 t
2. 6 t 773 kg = 6.773 t
3. 11 t 817 kg = 11.817 t
4. 7 t 995 kg = 7.995 t
5. 4 t 12 kg = 4.012 t
6. 6 t 4 kg = 6.004 t
7. 98.5°F
8. 99.9°F
9. 100.1°F
10. 26.5 kg
11. 26.1 kg
12. 26.6 kg

CHAPTER 4

Pages 26—27

1. 0.750 t
2. 0.900 t
3. 0.050 t
4. 1.250 t
5. 0.075 t
6. 70 kg
7. 1450 kg
8. 1750 kg
9. 2100 kg
10. 4650 kg
11. 200 kg = 0.200 t
12. 450 kg = 0.450 t
13. 750 kg = 0.750 t
14. 1250 kg = 1.250 t
15. 1700 kg = 1.700 t
16. 5.250 m, 5.037 m, 5.041 m
17. 15.328 m
18. 0.475 m, 0.262 m, 0.266 m
19. 5.504 t
20. 110 x 50 = 5500 kg = 5.500 t
21. No
23. 6.250 kg
22, 24, 25. See chart below. *is underweight delivery

14 April	9.792	4.288	5.504	110	5.500	
21 April	9.845	4.190	5.655	112	5.000	
27 April	10.673	4.165	6.508	130	6.500	
5 May	11.491	4.251	7.240	145	7.250	*
12 May	10.209	4.199	6.010	120	6.000	

Answers

1. 700 × 4.5 = 3150 ℓ
2. 8 × 0.3 = 2.4 m
3. 20 × 0.45 = 9 kg
4. 60 × 0.09 = 5.4 m^2
5. 3 × 0.6 = 1.8 ℓ
6. 150 mm
7. 180 m^2
8. 250 kg
9. 4.5 m
10. 4 ha
11. 2.1 oz
12. 140 mℓ
13. About 13 ft.
14. 280 g
15, 16, 17, 18, 19. correct
20. No. 60 ha ≈ 150 acres
21, 22, 23, 24. correct

Page 31

1. 4.73
   3.24
   4.73
   3.24
   ─────
   15.94 m ≈ 16 m

2. 16 × 3 = 48 m^2

3. 2 × 48 = 96 m^2
   so 6 ℓ is needed
   (6 × 17 = 102)

4. 7 ℓ of Polypaint needed so 3 tins required
   so cost = £7.02

5. Polypaint *much* better as 2 tins of Duracolour
   would be needed, costing £13.18

6. Area ≈ 150 m^2

7. 13 × 3 = 39 ℓ needed so get 8 large tins (£57.60),
   and total cost is £77.60

8. Best to *over*-estimate to be on the safe side

9. 168 m^2

10. 28 ℓ needed, so buy 30 ℓ at £28.08

11.
Hall walls	£ 32.95
Entrance	£ 7.02
Hall ceiling	£ 28.08
Hall floor	£ 77.60
	£145.65

12. So estimate £150.00

13. Paint could be spilt, walls and ceiling coverage
    could be different etc.

Page 32

1. 40
2. 40 × 13 = 520, so he can
3. 140 bags
4. 80000 m^2
5. 80 bags
6. 2 t
7. 2½ t
8. 3 bags
9. They need 83 (and 8 more spares) — total 91
   bags. So they have enough
10. 40000 m^2
11. 4 ha
12. 52 ha
13. 130 acres

CHAPTER 5

Page 35

1. 100 miles in 4 hours (25 mph)
2. 69 miles in 3 hours (23 mph)
3. 58 miles in 2 hours (29 mph)
4. 16 more miles
5. 32 mph
6. 180 ÷ 6 = 30 mph

7–10 Carol finds the *travel* time first.
   Total time = 22:00 − 8:30 = 13:30 or 13½ hours.
   Time *not* on road = 4 + ½ + 1 = 5½ hours at least.
   So longest travel time = 8 hours. So longest total
   distance = 8 × 35 = 280 miles. (140 miles out,
   140 miles back). So they can visit Cambridge
   (return at 19:00 hours) Scarborough (return at
   20:45 hours) and Bath (return at 22:00 hours)
   But Lake Windermere is too far.

Page 36

1. 3 mins.
2. 60 ÷ 4 = 15 mins.
3. 15 mins.
4. At 4 cars per min., 30 minutes
5. First car would leave 10 mins. after last one
   arrives, so space after 10 mins.
6. No. All figures are bound to be approximate
7. 22, 34, 50, 60 (assuming all cars stay ½ hour)
8. 11 (all cars arrived before 3:00)
9. 4 spaces filled between 3:20 and 3:25. 5 of next
   11 spaces filled between 3:25 and 3:30, so *6* spaces
   at 3:30 p.m.

Answers

1. 6.75 kg

2. Approx. 150 g, 125 g, 50 g (working to nearest 25 g)

3. Approx. 40 g per month

4. Yes, 1.5 kg per month is 4.5 kg per 3 months, which is more than 2.5 kg

5. 6 kg gain by roughly 4 months

6. 600 mℓ per day or 4200 mℓ per week

7. 68 beats per min.

8. 69 beats per min.

9. 114 beats per min.

10. 82 beats per min.

11. 14 beats per 10 sec. = 84 beats per min. A bit close — check again!

12. 96. Yes!

CHAPTER 6

Page 40—41

1. 62, 18, 8, 12, 48, 10, 6.

2. 100, 100

3. Low prices

4. 65%, 15%, 0%

5. 10%, 45%, 15%, 20%, 10%

6. 72, 60, 60%

7.

M	F
20	7
65	60
15	20
0	13

M	F
10	13
45	10
15	70
20	3
10	3

Pages 42—43

1. 6 x 0.85 = £5.10

2. 6 x 6.70 = £40.20

3. £9.15

4. £10.80

5. £67.50

6. 15 x 0.036 = £0.54 or 54p

7. £72.50

8. £13.60

9. £44

10. £73 + £17.52 ≈ £91

11. £57 (to nearest £)

12. £18.20, £72.80

13. Ken — £88.29
    Susan — £45.60

14. £11.64

15. Increase is £54
    10% of £180 is £18, 20% is £36
    so £54 is 30% of £180

Pages 44—45

1,2,3. See complete chart:

	Consumption			%
Month	1978	1979	Drop	Drop
Jan.	2750	2450	300	11
Feb.	2450	2350	100	4 *
Mar.	2600	2600	0	0
Apr.	2100	1950	150	7
May	1900	1850	50	3
June	1400	1500	−100	− 7

* Ron works out this % like this: 1% of 2450 = 24.50
  So 100 is about 4% of 2450.

4. 1978 — 13200 gal. 1979 — 12700 gal

5. Total drop = 500 gal

6. 132 gal

7. About 4% (3.8%)

8. 660 gal

9. 14600 gal

10. 14000 gal

11. 730 gal

12. Target — 13870. Actual — 14000.
    Yes (gap is smaller) 130

13. 16200, 15450. Target — 15390. Actual — 15450.
    Gap is now 60

14. Drop is 40. % drop = 6.4.

Page 46

1. 25% (¼)

2. 8.3 . . . %

3. 6.66 . . . ≈ 6.7%

4. £2.055 = £2.05½

Page 47

1. Correct

2. No. New price should be about £81 (12% of £72 = £8.64)

Answers

3. If increase is 20%, cars this year 8160 (6800 + 1360) per hour

4. 0.70 x 8 = 5.6p, so new price about 64.5p

5. After 1 year 3500 + 525 = £4025. After 2nd year 4025 + 604 = £4629

6. £49.64.

Page 48

1. 100 g

2. 210 calories

3. 120 g

4. 120 x 4.2 ≈ 500 calories

5. 1000 g

6. 700 calories

7. 375 g, say 380 g

8. 950 calories

9. Chicken for fewest calories — and it's not as expensive as beef! See 12.

10. 125 + 210 + 210 + 420 = 965 calories

11. 4 + 30 + 9 + 25 = 68 g of protein

12. Beef — 52p, chicken — 11p, milk — 30p, cheese — about 20p, bread — 15p

CHAPTER 7

Pages 50—51

1. Middle 'gross pay' should be £93.60

2. Net pay = 77.40 − 20.24 = £57.16

3. 8 hrs 38 min, 8 hr 51 min, 8 hr 30 min, 9 hr 27 min

4. 46 hr 00 min

5. 6 hours (46—40)

6,7. See wages slip below:

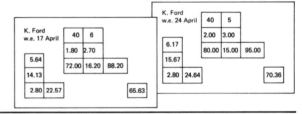

Pages 52—53

1. Yes (2.10 less than 2.16)

2. Yes (3.25 less than 3.57)

3. Yes (4.26 less than 4.32)

4. Yes (3.71 less than 3.91)

5. At least 6 of B

6. More than 25 of A

7. At least 9 of D

8. 4 hr + 4.25 hr = 8.25 hr

9. 1.25 hr

10. See chart below:

Employee: G. Dean1						Standard time (hr)	Bonus time (hr)	Bonus Pay (£)
Date	JOB							
	A 0.08 hr	B 0.36 hr	C 0.17 hr	D 0.24 hr				
Monday 7.8 st. time	50 4.00	—	25 4.25	—		8.25	1.25	2.50
Tuesday 8.8 st. time	40 3.20	—	31 8.37	—		11.57	4.57	9.14
Wed. 9.8 st. time	45 3.60	—	28 7.56	—		11.16	4.16	8.32
Thurs. 10.8 st. time	—	12 4.32	15 4.05	—		8.37	1.37	2.74
Friday 11.8 st. time	16 1.28	10 3.60	8 2.16	4 0.96		8.00	1.00	2.00

So bonus pay = £24.70.

11. 10 hr

12. 1.93 hr

13. Standard time = 8.93 hr. 37 x 0.24 = 8.88 hr 38 x 0.24 = 9.12 hr. So she must do at least 38 of Job D.

Pages 54—55

1. £4160

2. £2345

3. £187.50

4. £478.50

5. £666.00

6. Pay = £4680 per year. So taxable pay is 4680 − 1815 = £2865. Tax is   £187.50
                                          + £634.50
                                          £822.00

7.

8. £2185

9. £618 (187.50 + 430.50)

10. £11.88

11. Taxable pay = 5720 − 1440 − 1815 = £2465. So tax = £187.50
                        + £514.50
                        £702.00

12. £1939.60

13. £4.20

Answers

1. 36p

2. 200 ÷ 12 = 16.7 hours

3. Income = £10.50 Expenses: Stall £1.05
                                Materials £3.80
                                Bus   84
   So net income = £4.81

4. 69p per hour

5. £12.60 − £6.70 = £5.90

6. 74p per hour

CHAPTER 8

Pages 58—59

1. Wall loss   = £63
   Roof loss   = £45
   Draughts   = £27

2. Wall saving   = £38
   Roof saving   = £34
   Draught saving   = £16
   Windows   = £11
                      £99

3. Walls − 5 years (180 ÷ 38 = 4.74)
   Roof − 2 years (65 ÷ 34 = 1.91)
   Windows − 90 years (1000 ÷ 11 = 90 . . .)
   Draughts − 5 years (80 ÷ 16 = 5)

4. Interest on £65 (roof)   = £4.88 per year
                £180 (walls) = £13.5 per year
                £80 (draught) = £6 per year
   − less than savings in fuel costs.

5. Insulate roof!

6. 7 × 6.50 = £45.50. So taken over 2 years hire-purchase = £1092. So bank is cheaper

7. £798

8. Assuming they save £99 per year, they take 8 years to pay off their total costs.

Pages 60—61

1. About 250 therms

2. Estimate based on autumn, not winter usage

3. At 30 therms per week, say 400 therms

4. 20 × 50 = £10.00
   16 × 350 = £56.00
            £66.00

5. 2 × 66 + 2 × 33, say £200

6. No

7. Approx. £17

8. About 20 hcf (mid-way between 12 and 28)

9. 260 therms

10. £10 + £33.60 ≈ £44

11. 2 × 44 + 2 × 16 = £120

12. £10

13. 2000

14. About 2 years (666 days)

15. About 5 months.

Page 63

1, 2. See chart below:

	Power	Cost per Hour	Number of hours per week	Cost per week	Cost per quarter
Electric fire	2 kW	6p	30	£1.80	£23.40
Vac cleaner	300 W = 0.3 kW	0.9p	6	5.4p	£ 0.70
Kettle	3 kW	9p	2	18p	£ 2.34
100 W bulb	100 W = 0.1 kW	0.3p	40	12p	£ 1.56

3. 115 hours

4. 34½ p

5. Iron, convector heater

6. Chest freezer − £4.03
   Automatic washing machine − £2.21
   Colour TV − £1.56

Page 64

1. 4 legs, 2 sides, 2 ends

2. Slat + space between 95 and 105 mm
   12 slats = 900 mm
   11 spaces of ≈ 27 mm = 297 mm } = 1197 mm
   So 12 slats required

3. 600 mm + 100 mm + 100 mm = 800 mm

4. 2 × 1200 mm = 2400
   2 × 600 mm = 1200
   4 × 750 mm = 3000
   12 × 800 mm = 9600
              16200 ≈ 17 m

Answers

5. £5.10

6. Yes — 6 x 80 = £4.80

7. 12 x 18 = 216 tiles

8. 19 m^2

9. £108

10. 19 x £3.65 = £69.35
19 x £1.50 = £28.50
= £97.85

11. 900 mm, 900 mm

12. 14 blocks

13. 12 courses

14. 178 blocks.

CHAPTER 9

Pages 66—67

1. 5 kg

2. 40 kg

3. 5 kg

4. 8

5. 7 kg

6. 7, 56, 7 kg

7. 70 kg

8. 208 g

9. Water lost in cooking

10. 1

11. 40 g

12. 100 g

13. 5

14. 75 g

15. 300 g

16. 600 g

17. 800 g

18. 20 ml

19. 50 ml

20. $^1/_3$

21. 7.5 ℓ

Pages 68—69

1. 1.75 m^3

2. 1 m^3

3. 500 kg

4. 10 bags

5. 20, 40

6. 50, 100

7. 5 ℓ, 10 ℓ

8. 12.5 ℓ cement, 25 ℓ sand

9. 5 shovel-fuls cement, 10 shovel-fuls sand, 20 shovel-fuls gravel

10. 25 to 50 to 100 = 12½ to 25 to 50
= 5 to 10 to 20

11. 50 x 2.4 = 120 kg. 2.4 (approx. 2½) barrow loads

12.

Volume concrete	gravel		sand		cement	
	volume	shovels	volume	shovels	weight	shovels (bags)
50 ℓ (0.050 m³)	50 ℓ	20	25 ℓ	10	12.5 kg	5 (¼)
100 ℓ (0.100 m³)	100 ℓ	40	50 ℓ	20	25 kg	10 (½)
500 ℓ (0.5m³)	500 ℓ	200	250 ℓ	100	125 kg	50 (2½)
1000 ℓ (1 m³)	1000 ℓ	400	500 ℓ	200	250 kg	100 (5)
2 m³	2000 ℓ	800	1000 ℓ	400	500 kg	200 (10)

Pages 70—71

1. A — ½, B — 1$^3/_4$, C — 2$^3/_4$, D — 3$^3/_8$, E — 4$^3/_{16}$, F — 4$^5/_8$, G — 4$^{13}/_{16}$, H — 5½, I — 5$^{11}/_{16}$, J — 5$^7/_8$

2. $^3/_8$ = $^6/_{16}$, 1$^2/_4$ = 1½ = 1$^8/_{16}$

3. $^1/_{32}$'s, $^1/_{32}$'s

4. Fuel $^5/_{60}$; Housekeeping $^{28}/_{60}$; Travel $^7/_{60}$ Insurance $^2/_{60}$

5. $^5/_{60}$ = $^1/_{12}$, $^{28}/_{60}$ = $^{14}/_{30}$ = $^7/_{15}$, $^7/_{60}$, $^2/_{60}$ = $^1/_{30}$

6. $^5/_{60}$ + $^{28}/_{60}$ = $^{33}/_{60}$

7. $^{11}/_{20}$

8. $^9/_{60}$

9. $^{38}/_{70}$ = $^{19}/_{35}$

10. 15 ha

11. 3 ha

12. $^6/_{60}$ = $^1/_{10}$

13. 35 ha

14. $^1/_3$

15. ¼

16. $^2/_3$

17. $^5/_6$

18. $^1/_5$

19. ¾

20. $^4/_5$

21. $^1/_{20}$

22. 6

23. 18

24. 9

25. 45

26. 9

27. 27

28. 2.5 or 2½

29. 12.5 or 12½.

Page 72

1. 1.4 m

2. 4.23 m

3. 7.06 m

4. 2.381 m

5. $^{75}/_{100}$ or 75%

6. $^{21}/_{100}$ or 21%

7. $^8/_{100}$ or 8%

8. $^3/_{10}$ = $^{30}/_{100}$ or 30%

9. 0.5, 0.25, 0.0625

10. 0.75, 0.875, 0.625, 0.1875

11. 0.27 x 61 = 16.47 m

Answers

CHAPTER 10

Page 75

1, 2. See table below:

Measurement	Fullsize (cm)	On Plan cm	On Plan No. squares
Length	260	13	26
Width	230	11.5	23
Window — A	60	3	6
Window width	130	6.5	13
Window — B	70	3.5	7
b. door — A	140	7	14
b. door width	70	3.5	7
h. door — C	60	3	6
h. door — width	60	3	6

3.

60 mm

30 mm — Double storage unit with worktop

20 mm — hinge — 30 mm — snack flap

27.5 mm — refrigerator — 30 mm

Page 76

1. 10 mm
to B — 65 mm = 6.5 m
to A — 34 mm = 3.4 m
GH — 26 mm = 2.6 m
FE — 111 mm = 11.1 m
to C — 31 mm = 3.1 m
DE — 7 mm = 0.7 m
E I — 49 mm = 4.9 m

3. 10.85 x 4.65 m

4. Approx. 50 m^2

5. 77 + 30 = 107 m^2 so extra area = 57 m^2

6. The cost is for the whole floor area so likely charge will be around £170 x 107 = £18190

Page 77

1.

Full size	40 m	30 m	6 m	20 m	5 m	18 m
Scale (1 to 400)	100 mm	75 mm	15 mm	50 mm	12.5 mm	45 mm

2. See plan; Note — the plan is half-size, scale 1:800.

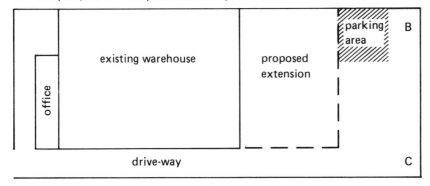

7. Two lorries can just pass in the driveway, and at B, or the lorry parking area. (Full-size lorry measurements are 12 m by 2.4 m width.)

Page 78

1. 250 m

2. 4 cm

3. 1⅝ km (about 6½ cm)

4. 5½ km (about 22 cm) (Ending where you began)

5. 1 hr 22½ mins

6. 1¼ km (5 cm)

7. About 3.30 p.m.

8. 75 m (3 mm x 25000 = 75000 mm)

9. 2 cm

10. 325 m (13 mm x 25000 = 325000 mm)

11. 150 m

12. 650 m

13. 240 m

14. 15 mm

15. 80 cm

16. 12.8 km ÷ 25000 = 12.8 m ÷ 25 = 51.2 cm

Answers

CHAPTER 11

Page 80

1. 700 g

2. 10½ oz, 28½ oz

3,4

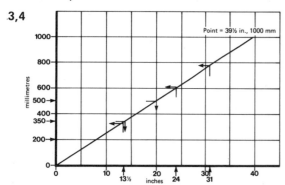

5. 24 ins = 610 mm,
   31 ins = 780 mm
   13½ ins = 340 mm

6. 200 mm = 8.1 ins
   350 mm = 14 ins
   500 mm = 20 ins

7

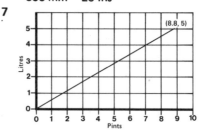

Page 81

1. 14 lb, 6.5 kg. 24 lb, 11 kg

2. 1 to 6 months

3. 5.5 kg roughly

4. 11.5 − 7 = 4.5 kg

5. No, if anything underweight

6. 10 stone

7. About 11 stone

8. Between ½ and 2 stone. Say 1 or 1½ stone.

Page 82

1. 14 min 2. 5 min 3. 15 min 4. 13 min 5. Just after 0812 6. (a) and (d).

Page 83

1. About 5 years

2. £1 becomes £1.80 in 6 years. So £100 becomes £180 and £200 becomes £360

3. About 62p

4. Just below 10% line. So about 11%

TEST 1

1. 2 x 30p, 4 x 30p, 1 x 40p, 3 x 30p, 2 x 20p, 1 x £4.30, 1 x £1.70 and 1 x £3.90

2. £13.40

3. Yes, approx. £3.50

4. £49.70

5. £13.34, £49.61

6. Plane   £11
   Screwdriver   £ 0
   Clamp   £ 1
   Drill   £ 3

Tape measure —   £ 1
Spirit level —   £ 4
Wood-saw —   £ 4
Hack-saw —   £ 3
Brace —   £ 6
Hammer —   £ 2
Set Square —   £10

7. £45

8. No

9. 8, everything except the coffee

10. 52p.

TEST 2

1. Yes (8 + 15 − 18 = 5 — the amount at the end of the week)

2. 6   4. 180   6. £204.80.

3. 53   5. 1024

TEST 3

1. 36.7   6. 18.5   11. millimetres   16. millilitre

2. 37.6   7. 19.7   12. ml   17. m²

3. 38.05   8. 20.5   13. centimetres   18. False, false, true.

4. 5.05   9. 22.2   14. 1000

5. 5.37   10. metres 15. millimetres

Answers

TEST 4

1. 2400 kg
2. 360 kg
3. About 7 m
4. Approx. 30 days
5. 2135 kg, 3160 kg, 4000 kg
6. 340 pints = 42.5 gallons. About 9 oz.

TEST 5

1. Medium
2. 3 mℓ
3. Large
4. 2.3 mℓ per penny
5. Slightly better
6. 2 mℓ per p.
7. 5p
8. £5.20
9. 14 shampoos, 7 weeks
10. £5.20 − £4.46 = 74p.

TEST 6

1. 60p
2. £11.40
3. £25.60
4. 88p
5. 15%
6. 33%
7. 86%
8. 56%
9. £12.60
10. £294.40
11. £13.86
12. £51.75
13. Yes
14. No
15. Yes
16. Yes
17. 6%
18. 12%
19. £1.84
20. £105.74

TEST 7

1. £101.92
2. £847.50
3. £70.80
4. $150 \div 4.4 \approx 34$ weeks.

TEST 8

1. £175
2. About £149
3. 4.5p, 2.4p
4. 42p
5. About £5
6. £56
7. £84
8. £52
9. About 2½ years.

TEST 9

1. 5.1 m
2. 5.1 metres per second
3. 11.22 mph
4. 1½ turns per second
5. 4, 4, 24
6. $^6/_8$ or ¾
7. All equal to ¼ except $^4/_{10}$
8. 375 g, 150 g, 250 mℓ, 50 g

TEST 10

1. 25000 cm
2. 250 m
3. 4 cm
4. 11.5 cm
5. 3 km
6. ¾ hr (45 mins.) approx.
7. 1 to 25
8. See scale drawing →
9. 1.150 m.

Index

MATHS INDEX — use this if you want to find where a mathematics skill is used. The arrows △ , ▽ will also help you find it.
S = look in Situation index below

SITUATION INDEX — look here to find where the job or situation you are interested in appears. But remember, the same maths idea can appear in lots of different places, so use this as a guide only. **M** = look in maths index above.